EUCALYPTS

A BUSHWALKER'S GUIDE

BUSH
BOOKS

SYDNEY
& ENVIRONS

EUCALYPTS

A BUSHWALKER'S GUIDE

GARY LEONARD

Illustrations by Suzanne Fyfe

NEW SOUTH WALES
UNIVERSITY PRESS

The author would like to thank Chris McTaggart, Leon Fuller,
Dr Kevin Mills, Arnold Vink, Rob Attwood, John Rowell,
Dr Ken Hill and Dr Roslyn Muston.

Special thanks are due to
Van Klaphake and Suzanne Fyfe.

BUSHBOOK TITLES

Sydney
and
environs series

EUCALYPTS: A Bushwalker's Guide

SEASHORES: A Beachcomber's Guide

Published in Australia by:
NEW SOUTH WALES UNIVERSITY PRESS
PO Box 1 Kensington NSW Australia 2033
Phone (02) 398 8900 Fax (02) 398 3408

© Gary Leonard
First published 1993

National Library of Australia
Cataloguing-in-Publication entry:

Leonard, Gary, 1947–
    Eucalypts: a bushwalkers guide.

    ISBN 0 86840 340 7.
    ISBN 0 86840 348 2 (series).

    1. Eucalyptus – New South Wales – Newcastle Region –
    Identification. 2. Eucalyptus – New South Wales – Wollongong
    Region – Identification. 3. New South Wales – Guidebooks. I.
    Fyfe, Suzanne, 1963-   . II. Title. (Series: Bush books
    (Kensington, N.S.W.).)

583.42099442

Available in North America through:
I.S.B.S. Inc
5804 N.E. Hassalo St.
Portland, OR 97213-3644
Tel: (503) 287 3093
Fax: (503) 280 8832

CONTENTS

HOW TO IDENTIFY THAT EUCALYPT

The identification of eucalypt species can be a daunting task, even for those with some botanical knowledge. Traditionally, botanical keys are used, but these require a reasonable knowledge of plant morphology. This field guide requires no such background knowledge and is designed for people who want to identify eucalypts, either to decide which species to grow in their gardens or simply to develop a better understanding of their local vegetation.

This field guide describes 68 species in the genus *Eucalyptus* commonly found growing between Newcastle in the north, Katoomba in the west, and Wollongong in the south. Only species that are native to this area are described.

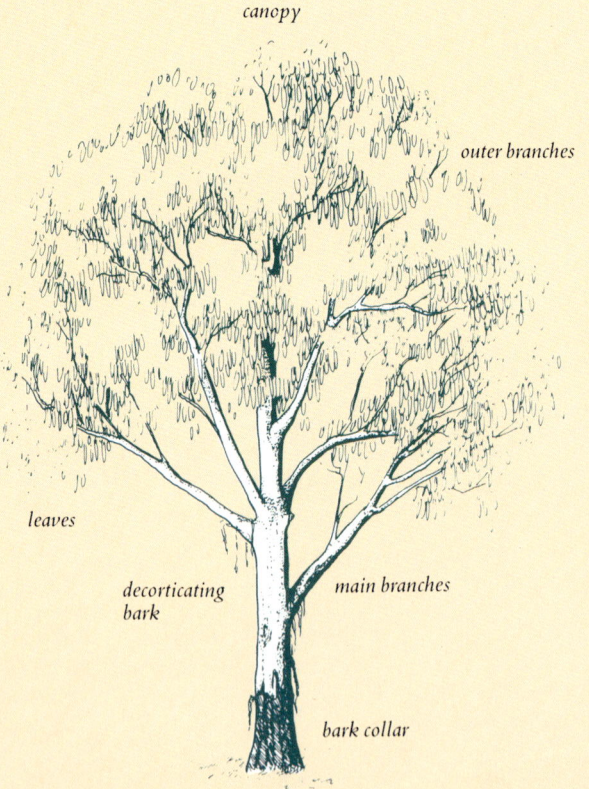

canopy

outer branches

leaves

decorticating bark

main branches

bark collar

A typical tree

FIRST STEPS

Most people will have no difficulty recognising a eucalypt, but if there is any doubt the following checks could be carried out.

1. Crush a leaf for the characteristic smell of eucalypts.

2. Look for woody capsules; or flower buds with 'caps'.

3. The leaves should be thick, tough and arranged alternately along the stem. (Juvenile leaves are often arranged oppositely, but these leaves are usually softer, bluish and usually only appear at the ends of the stems.)

4. Most eucalypts have a more open appearance than other trees. Eucalypts do not screen a view so much as soften it.

5. Growth habit may vary from a single erect trunk, to a many-trunked clump, referred to as a 'mallee'.

6. Bark type varies, but in many species the bark is deciduous, peeling away in plates, patches or strips.

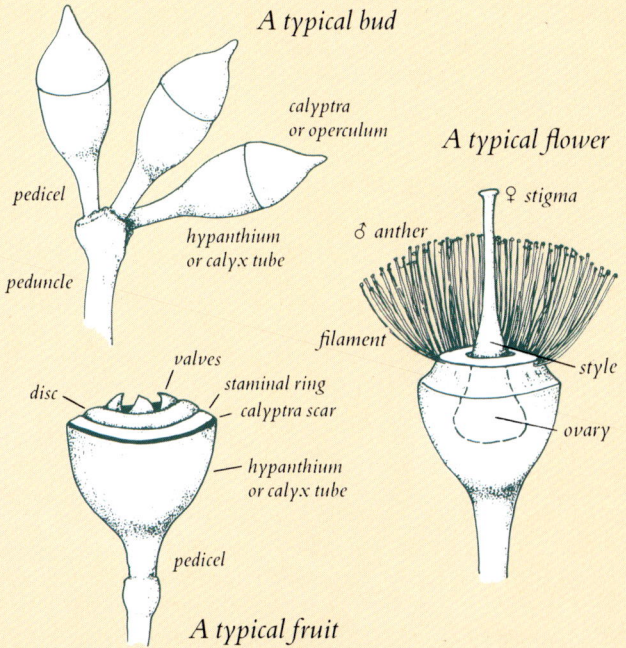

A typical bud

calyptra
or operculum

A typical flower

pedicel

♂ anther

♀ stigma

hypanthium
or calyx tube

peduncle

filament

valves

style

disc

staminal ring

calyptra scar

ovary

hypanthium
or calyx tube

pedicel

A typical fruit

THE NATURAL HABITAT

Once we have confirmed that the tree is a eucalypt, we need not immediately examine the tree in detail. We may find useful clues by observing where the tree is growing.

Many species grow only in certain habitats. For example, the mountain grey gum and the gully gum grow only at altitudes higher than 300 metres above sea level, whereas the cabbage gum and the river peppermint favour creek and river banks. By identifying the habitat we automatically cancel out many species.

Woodland on Hawkesbury Sandstone.

SOILS

Soil type may also restrict many species. Ironbarks tend to prefer clay soils, whereas Sydney peppermint and silvertop ash favour Hawkesbury Sandstone.

Hawkesbury Sandstone soils are derived from the major rock type in the Sydney basin – a feature of this rock type being the spectacular cliff formations along much of Sydney Harbour shores, the coastline of the Royal National Park and the top layer of the Illawarra escarpment. Soils derived from this rock type tend to have a low amount of nutrients considered necessary for good plant growth, although many native plants have evolved methods of overcoming this 'deficiency'. The soils tend to be coarse, generally pale yellow to light brown in colour, and usually contain quartz particles. Drainage is often a problem because of the impervious nature of the underlying sandstone. In low-lying areas, swamps and bogs are often formed.

The Wianamatta group consists of a series of shales and sandstones that overlie Hawkesbury Sandstone in large parts of western Sydney, Glenbrook in the Blue Mountains, the Moss Vale district, and Darkes Forest (south of Helensburgh). This material forms finely textured red-brown soils, which are quite fertile compared with Hawkesbury Sandstone soils. Ashfield Shale is derived from this group and is the most frequently experienced in the Sydney basin.

The Narrabeen group of rocks forms the more gentle slopes of the Illawarra escarpment, immediately below the steeper Hawkesbury Sandstone cliffs. They form the talus slopes that support rainforest and tall eucalypt forests, as well as providing rich alluvial soils for the flood plains below. They may be also seen in the steeper cliffs along the Hawkesbury River. Soils derived from the Narrabeen group are finely textured clays, often with high nutrient levels, varying in colour from grey to red to chocolate-brown.

The Illawarra Coal Measures, apart from providing coal, produce a dark brown, reasonably fertile clay soil. This rock type forms a series of foothills along the base of the Illawarra escarpment.

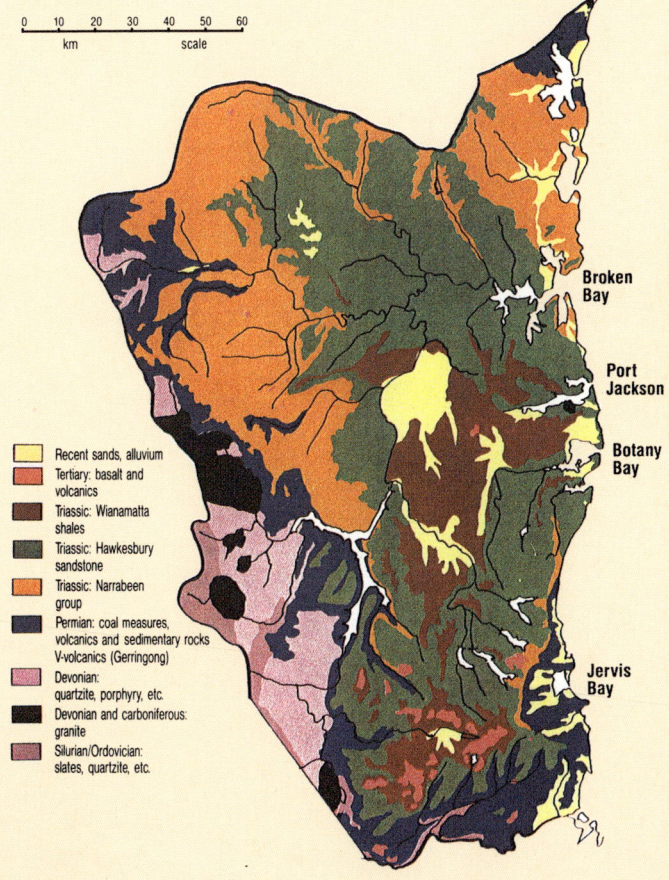

Soil map of the Sydney region (from Field Guide to the Plants of the Sydney Region *by Fairley and Moore, published by Kangaroo Press, Sydney).*

NATIONAL PARKS

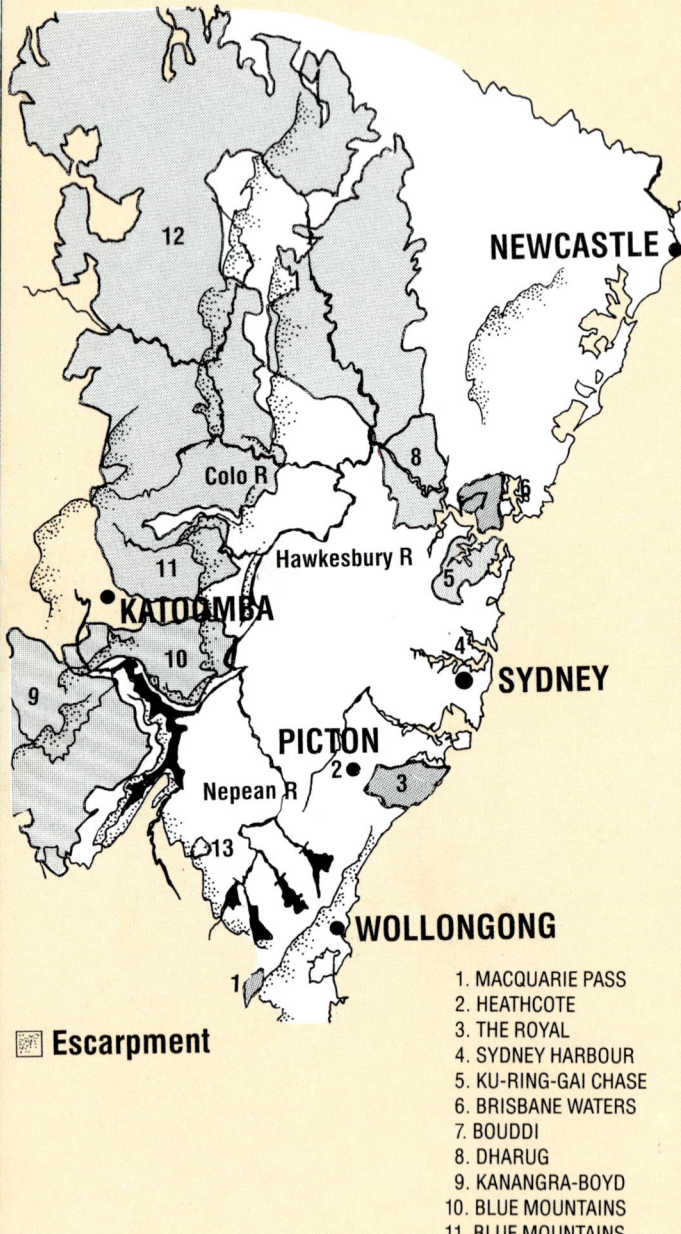

Escarpment

1. MACQUARIE PASS
2. HEATHCOTE
3. THE ROYAL
4. SYDNEY HARBOUR
5. KU-RING-GAI CHASE
6. BRISBANE WATERS
7. BOUDDI
8. DHARUG
9. KANANGRA-BOYD
10. BLUE MOUNTAINS
11. BLUE MOUNTAINS
12. WOLLEMI
13. THIRLMERE LAKES

ASSOCIATED SPECIES

In any given habitat, most eucalypt species grow in association with other eucalypt species. For example, red bloodwood grows with scribbly gum on Hawkesbury Sandstone. If you recognise one species, look it up in the book to get the names of associated species, then refer to the pages for each of them to check their identity.

Monotypic stands do occur, however. You may find large areas of the spotted gum, with no other eucalypt species in sight.

Typical open forest.

Some eucalypt species grow only in one community of plants. Others may appear in entirely different forms if they are able to grow in different communities. Mallees are small and shrubby, with several trunks, and are restricted to wood-

Monotypic stand of spotted gum with understorey of Macrozamia sp.

Angophora costata.

land; trees in woodland tend to have low spreading branches and are often twisted or leaning. In contrast, trees in open forest have tall, straight trunks and ascending branches, and are spaced more closely than in a woodland. If the trees exceed 30 metres in height, the community is called a tall open forest.

Now Take A Close Look

A key to different types of bark is set out below. Find the description that seems closest to the bark of your tree, then turn to the species number to check the identification. With each species description there is a small diagram illustrating typical tree form. If the bark decorticates leaving a collar then an arrow roughly indicates end of the collar.

KEY TO BARK		Species No.
IRONBARK Bark deeply furrowed; very hard; corky when old	Covering all branches	62, 63, 64, 65, 66, 67
	Covering only larger branches; upper branches smooth	19
STRINGYBARK Bark may be pulled off in long fibrous strips	Covering all branches Fine	4
	Coarse	5, 6, 7, 8, 9, 10, 12, 14, 15
	Scaly	16
	Covering only larger branches; upper branches smooth	11, 13
STRINGYBARK-LIKE Soft, fibrous, but can't be pulled off in long strips	Covering only larger branches; upper branches smooth	68
	Covering all branches	17
BLOODWOOD BARK Bark flaky; can be pulled off in chunks	Covering all branches Chunky brown–red Flaky brown–yellow	1 2
PEPPERMINT BARK Bark thin, finely fibrous; not stringy or furrowed; peels off in fibrous sheets	Covering all branches	32
	Covering only larger branches; upper branches smooth	20, 33
BOX BARK Bark thin, finely furrowed; can't be pulled away from trunk	Covering all branches	57, 58
	Covering trunk only; all branches smooth	59, 60, 61
ROUGH, VARIABLE COLLAR with smooth upper trunk and branches	Grey collar, cream branches	13, 51
	Brown collar, white branches	33, 53, 54, 55
	Mottled trunk and branches	44, 45, 46, 47
ROUGH, VARIABLE COLLAR to larger branches; upper branches smooth	Brown collar, white branches	18
	Grey collar, cream branches	19, 50
	Grey collar, grey branches	42
	Flaky bark on collar	50, 37, 52
	Soft, furrowed bark on collar	35, 36, 38, 41, 42
BARK SMOOTH; NO COLLAR	Mottled trunk and branches	3, 48, 49
	Mallee	21, 22, 23, 24, 25
	Scribbles on trunk	26, 27, 29, 30, 31
SHORT COLLAR; MAINLY SMOOTH TRUNK	Mallee	28
	Branches white	29, 56
	Branches cream to grey	34, 36, 37, 39, 40, 42, 55
	Branches mottled	52, 53, 54

Now that you've checked the details of a species' habitat, soil preference, community and association, and studied the tree's form and its bark, you'll be able to identify what species it is by examining its leaves; the buds and flowers; and the fruit.

13

CONFUSING FACTORS IN IDENTIFICATION

Individual trees of some species of eucalypts may vary markedly in appearance. Here are some reasons.

CONVERGENCE

Trees will adapt to different conditions. For example, woollybutt growing with blackbutts will appear reasonably similar to the blackbutts, while one kilometre away the same species may have darker, thicker bark and be growing in association with grey ironbarks.

HYBRIDIZATION

Related species may interbreed, especially in disturbed areas. For example, between the Georges River and Nowra, bangalay and Sydney blue gum hybridize, resulting in trees with characteristics of both species. Many ashes and stringy-barks may be difficult to identify as a result of hybridization.

JUVENILITY

The foliage on young trees, and even new growth on old trees, may appear quite different from what you might expect. This is because most eucalypts produce juvenile and intermediate foliage before adult leaves appear. The juvenile leaves are arranged opposite each other on the stem and tend to be larger and a different colour from adult leaves. Juvenile

Epicormic shoots after a bushfire.

leaves are not described in this field guide, so please take care that your leaf samples are from mature foliage. The leaves should be arranged alternately on the stem, and if buds or fruit occur you can assume that your foliage is mature. Juvenile foliage will be particularly noticeable after a bushfire, when damaged trees attempt to recover by means of lignotuberous growth, from the roots, or epicormic growth, from the trunk or branches.

ANGOPHORA AND SYNCARPIA

You should be careful not to confuse species such as *Angophora* and *Syncarpia* which while similar are not true eucalypts. Angophora is closely related to eucalypt, but differs in the following ways: Angophora has only adult foliage; eucalypt

Angophora costata.

has juvenile. Angophora leaves are opposite; eucalypt leaves are alternate. Angophora seeds are larger than eucalypt seeds. Angophora flower buds do not have an operculum. Angophora capsules have ribs; most eucalypt capsules are smooth.

Angophora costata grows on Hawkesbury Sandstone, and is common on the coastal strip between Newcastle and Sydney, in Ku-ring-gai Chase National Park and Royal National Park, and along the northern cliffs of Sydney Harbour. It is found growing in association with Sydney peppermint, red bloodwood or scribbly gum, and sometimes with an understorey of waratah and other leathery-leafed shrubs. The bark is smooth, mottled red to grey, and peels away in patches, leaving dimples on the trunk. The branches are often twisted and broadly spreading.

Angophora hispida is usually found as a low, spreading shrub on Hawkesbury Sandstone in heath. The leaves are broad and hairy, and in summer the large heads of white flowers make this a spectacular plant. Good specimens may be seen along Mona Vale Road and Heathcote Road.

Angophora floribunda has rough bark and favours undulating country and flood plains, growing in association with forest red gum, woollybutt, white stringybark and grey gum. The bark is grey, and rough on all but the smallest branches.

Angophora bakeri also has rough bark, but may be distinguished by its narrow leaves. This species is mostly found on Hawkesbury Sandstone.

Angophora subvelutina, another rough-barked species, is often found growing with grey box, especially along the Nepean River. Like *A. hispida*, this species has leaves with a cordate (heart-shaped) base.

Syncarpia glomulifera, turpentine, is also in the same botanical family as *Eucalyptus* and may be found growing on the fringes of the rainforest, in association with blackbutt, grey ironbark or Sydney blue gum. It has a brown 'stringy' bark but can be easily distinguished by its woody fruits, formed from seven fused flowers, and by its leathery leaves, which are dark green above and whitish beneath.

BLOODWOODS

Bloodwoods differ floristically from other eucalypts and eventually botanists will give them the status of a separate genus, Corymbia. Most bloodwood species are common trees, and they are not usually commercially logged (a notable exception being spotted gum). The timber has veins and pockets of gum, making it difficult to work.

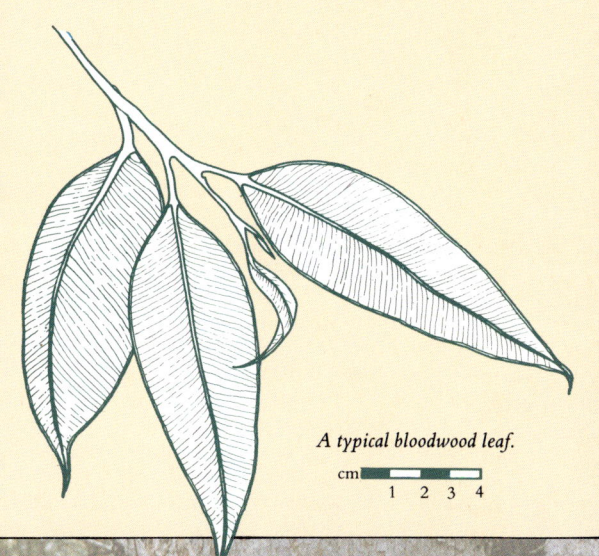

A typical bloodwood leaf.

cm ▮ 1 2 3 4

1

Red bloodwood
Eucalyptus gummifera

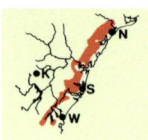

HABITAT Very common and widely distributed on hind-dunes and coastal plains, low hills and ridges.
SOILS Hawkesbury Sandstone and similar sandy soils.
COMMUNITY Open forest; woodland.
ASSOCIATED SPECIES Silvertop ash, scribbly gum, Sydney peppermint.
FORM Mallee on Hawkesbury Sandstone ridges; medium tree with large dense canopy in better conditions.
BARK Rough bloodwood-type bark, to small branches; grey to brown, often blackened by fire; tessellated.
LEAVES Glossy dark green above, much lighter beneath; veins regularly spaced and roughly parallel.
INFLORESCENCE 3 to 7; large, mostly at ends of branches, in late summer.
FRUIT Large, urn-shaped.

The wood is very strong and durable, making it a popular timber for sleepers and poles.

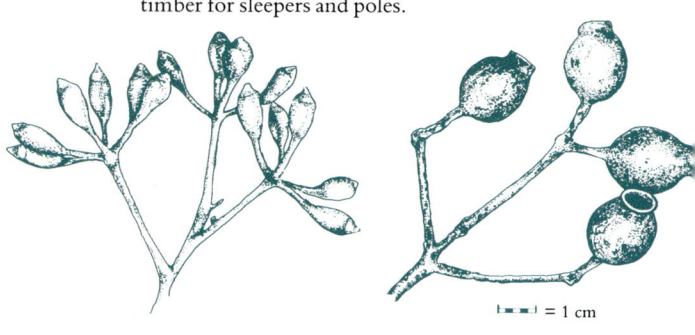

└──┴──┘ = 1 cm

2

Yellow bloodwood
Eucalyptus eximia

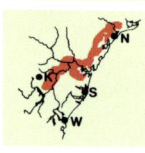

HABITAT Coastal hills and ridges.
SOILS Hawkesbury and other sandstones.
COMMUNITY Woodland; open forest.
ASSOCIATED SPECIES Scribbly gum, red bloodwood, Sydney peppermint.
FORM Small tree, straight short trunk, large dense clumps of pendulous foliage; open crown.
BARK Rough, to small branches; more flaky than red bloodwood; tessellated; brown, but new bark yellow.
LEAVES Large, pendulous, with conspicuous yellow midrib; new leaves purple, turning blue-green and leathery.
INFLORESCENCE 3 to 7; large cream flowers in spring.
FRUIT Urn-shaped.

This is a useful ornamental tree for areas such as Balmain, where shallow soil and small gardens are limiting factors. It is not common in the Sydney area, but stands may be found in Galston Gorge and Ku-ring-gai Chase National Park.

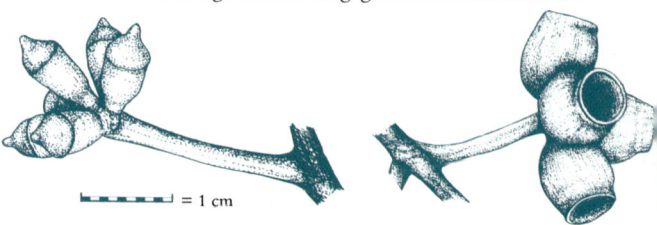

└──┴──┴──┘ = 1 cm

19

3

Spotted gum
Eucalyptus maculata

HABITAT   Coastal plains, valley slopes and moist ridges.
SOILS   Volcanic sandstones; shales.
COMMUNITY   Tall open forest.
ASSOCIATED SPECIES   Blackbutt, Sydney blue gum, grey gum, grey ironbark, narrow-leaved ironbark. Often seen in pure stands.
FORM   Long, straight trunk, narrow canopy.
BARK   Smooth; dimpled; no collar; bark shed in plates, resulting in a mottled pattern of white, pink and blue-grey.
LEAVES   Long and leathery; sometimes attacked by white lac lerp.
INFLORESCENCE   Usually in threes; winter to spring.
FRUIT   Urn-shaped to ovoid.

Eucalyptus maculata may be mistaken for *E. citriodora* in a garden situation; however, the lemon-scented gum can be distinguished by the leaves, which are narrower and strongly aromatic. It is native to coastal Queensland but a popular ornamental in Sydney gardens. The timber is in great demand for tool handles, flooring and poles.

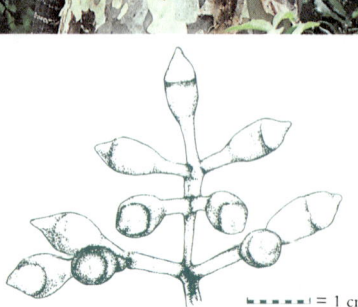

= 1 cm

STRINGYBARKS

The stringybark group may often be confusing and difficult to identify. Species are inclined to hybridize, especially in disturbed areas. They appear to hybridize not only amongst themselves but also with blackbutts, peppermints and ashes.

Most stringybarks can be recognised by their bark: rough, at least on the trunk, and able to be pulled from the trunk in long strings. Many species have leaves with an oblique base, and fruit are generally borne in tight clusters.

A typical stringybark leaf.

cm 1 2 3 4

4

White mahogany (yellow stringy-bark)
Eucalyptus acmenoides

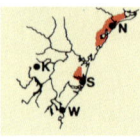

Not a true stringybark.

HABITAT   Dry coastal ridges and hills north of Sydney, although small stands may be found around Ryde, Galston and Parramatta.

SOILS   Shales.

COMMUNITY   Open forest.

ASSOCIATED SPECIES   Grey gum, grey ironbark, stringybarks, forest red gum.

FORM   Usually short thick trunk and open crown; in better conditions the trunk may be tall and the canopy quite dense, as is the case when growing in association with tallowwood.

BARK   Rough, to small branches; grey-brown; finely fibrous not as coarse as that of true stringybarks.

LEAVES   Lighter below; thin, finely tapering.

INFLORESCENCE   9 to 15; flattened peduncle; late spring to summer.

FRUIT   Hemispherical; valves level with rim; not as tightly clustered as true stringybarks.

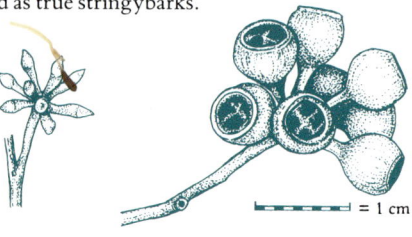

= 1 cm

5

Yellow stringybark
Eucalyptus muelleriana

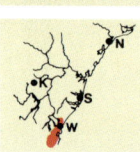

HABITAT   Sheltered sites, moist coastal hills and valleys, plains with deep clay soils, from Wollongong southwards.

SOILS   Narrabeen and other shales; alluvium.

COMMUNITY   Tall open forest.

ASSOCIATED SPECIES   Gully gum, mountain grey gum, coast white box, messmate stringybark.

FORM   Tall tree, often with very thick trunk and dense crown strongly spreading, straight branches.

BARK   Thick, long-fibred stringybark, to small branches; brown to grey, finer than most stringybarks.

LEAVES   Bright glossy green, only slight colour difference between top and bottom; oblique base.

SIMILAR SPECIES   Messmate does not have such stringy bark, but otherwise looks very similar when growing in association with yellow stringybark.

INFLORESCENCE   7 to 12; flattened peduncles; summer.

FRUIT   Hemispherical; valves level or slightly exserted; not rightly clustered.

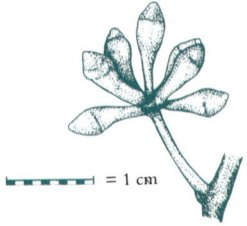

= 1 cm

23

6

Thin-leaved stringybark
Eucalyptus eugenioides

HABITAT   Common on coastal plains, Cumberland Plain (Parramatta and further west, between Hawkesbury River and Campbelltown) and lower Blue Mountains, especially on shale derived or alluvial soils. This tree is easily confused with the white stringybark but tends to favour drier conditions. The two species commonly hybridize in disturbed areas or where their distributions meet and overlap.

SOILS   Moderately fertile clays.

COMMUNITY   Open forest.

ASSOCIATED SPECIES   Forest red gum, coastal grey box, woollybutt, grey gum.

FORM   Short straight trunk; spreading, sparse crown, low-branching; may have tall trunk and narrow crown in forest situations.

BARK   Grey to brown; typical stringybark.

LEAVES   Green, almost no colour difference between top and bottom; oblique base; curved, thin and narrow (<25 mm).

INFLORESCENCE   5 to 12; peduncles slightly flattened; winter to spring.

FRUIT   Short stem; hemispherical; valves level with rim.

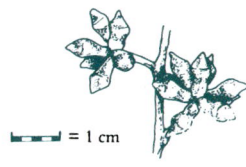

= 1 cm

7

White stringybark
Eucalyptus globoidea

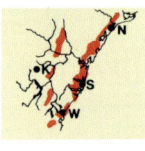

The white stringybark differs from the thin-leaved stringybark in the following ways:

HABITAT   Slightly more fertile and moist soils.

LEAVES   Thicker, shorter, broader (>25 mm); greater colour difference between top and bottom; juvenile leaves are hairier.

FRUIT   Smaller; very tightly clustered; mostly stalkless.

= 1 cm

25

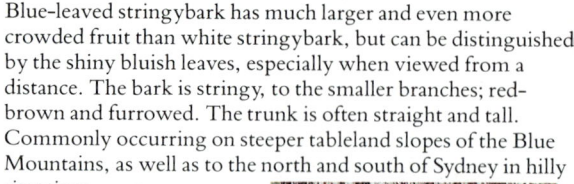

8

Blue-leaved stringybark
Eucalyptus agglomerata

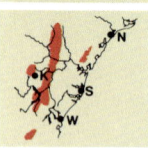

Blue-leaved stringybark has much larger and even more crowded fruit than white stringybark, but can be distinguished by the shiny bluish leaves, especially when viewed from a distance. The bark is stringy, to the smaller branches; red-brown and furrowed. The trunk is often straight and tall. Commonly occurring on steeper tableland slopes of the Blue Mountains, as well as to the north and south of Sydney in hilly situations.

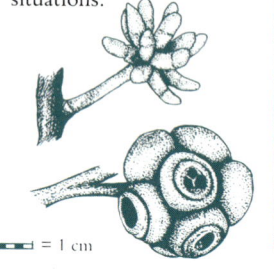

= 1 cm

9

Brown stringybark
Eucalyptus capitellata

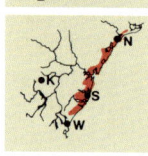

This small stringybark is mostly restricted to a small coastal strip north of Sydney and, in the south, in the Royal National Park. In sheltered situations this species may reach 15 m but is usually stunted. The bark is stringy, to the smaller branches; red-grey and furrowed. The fruit are crowded, their sides flattened from being pressed so closely together. The leaves are very thick, >25 mm in width, green and shiny top and bottom.

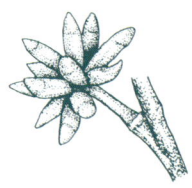

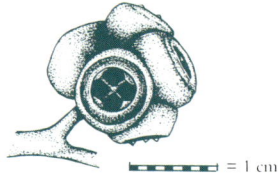

= 1 cm

10

Camfield's stringybark
Eucalyptus camfieldii

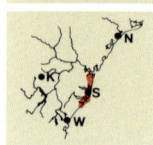

A small tree or mallee, found in isolated pockets, usually ridges of skeletal Hawkesbury Sandstone or lateritic soils in Mosman, Northbridge, Ku-ring-gai Chase, Heathcote and the Royal National Park. Associated species are scribbly gum and narrow-leaved stringybark. The bark is typically stringybark, and the fruit are crowded, but one distinguishing feature is the shape of the juvenile leaves, which are almost circular. Adult leaves are broad and thick; green and shiny top and bottom. The fruit are large, with a red rim.

This is an endangered species as a result of urban development and frequency of fires.

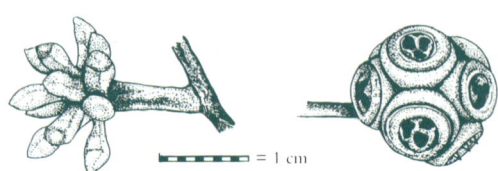

= 1 cm

11

Blaxland's stringybark
Eucalyptus blaxlandii

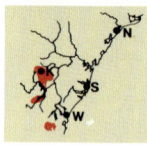

Commonly occurring small to medium tree of the upper Blue Mountains. The trunk has stringy bark, but the branches are smooth. The fruit are crowded and have a broad red disc and long exserted valves. The leaves are slightly glossy and the same colour top and bottom.

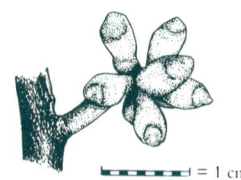
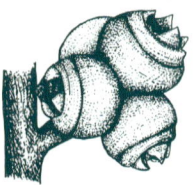

= 1 cm

12

Narrow-leaved stringybark
Eucalyptus oblonga

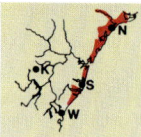

Small tree of coastal areas north and south of Sydney, usually o sandstone. The bark is rough and red-brown, to the small branches. The leaves are small, narrow and sometimes curved; glossy green top and bottom; thick. The fruit are crowded, the valves are level with the rim, the disc is broad and the orifice is small.

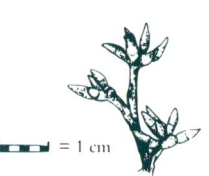
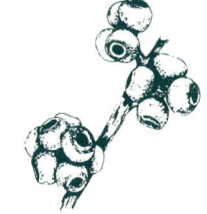

= 1 cm

13

Blackbutt
Eucalyptus pilularis

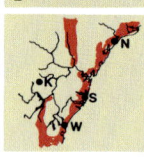

HABITAT   Coastal, common on lower slopes of Blue Mountains and Illawarra escarpment, and north of Sydney; fringes of rainforest; valleys; hind-dunes.
SOILS   Sandy loams; shales; some volcanic soils.
COMMUNITY   Tall open forest.
ASSOCIATED SPECIES   Sydney blue gum, grey ironbark, spotted gum.
FORM   Usually with tall, straight trunk; branches ascending.
BARK   Usually rough for a few metres from the base, but ver variable in extent and texture; grey, sometimes fire-blackened. Upper trunk and branches sometimes smooth; white or cream to yellow-grey.
LEAVES   Glossy green top and bottom; thick; curved.
INFLORESCENCE   7 to 15 flowered; summer; peduncle long and flattened or angular.
FRUIT   Hemispherical; valves enclosed.
SIMILAR SPECIES   May sometimes resemble Sydney blue gum and Sydney peppermint but can generally be distinguished by the strong V-shape of the branches and their whitish colour.

Blackbutt is one of Australia's most important hardwoods.

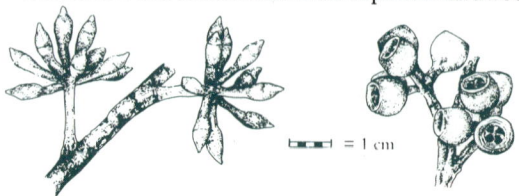

= 1 cm

14

Privet-leaved stringybark
Eucalyptus ligustrina

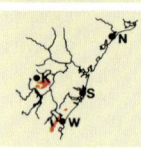

This species grows as a shrub, mallee or small tree along the edge of the Illawarra escarpment and in the Blue Mountains near Wentworth Falls. The bark is stringy on the main stems, but falling off in flakes on the smaller branches. Associated species are red bloodwood, silvertop ash and scribbly gum. Leaves vary in size from small to very small (40–60 mm long), glossy green on top, slightly paler below; usually curved. The fruit are crowded, with a small orifice, and valves inserted.

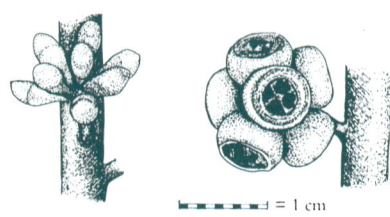

= 1 cm

15

Bastard mahogany
Eucalyptus umbra **subspecies** *umbra*

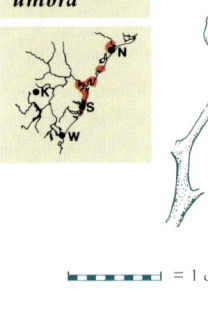

HABITAT   Poorer areas of the Hawkesbury Sandstone, from the northern shore of Sydney Harbour to Broken Bay, then along the coast to Newcastle.

MAIN DISTINGUISHING FEATURES   Stunted tree, with rough, short-fibred stringy bark, to smaller branches. The leaves are curved, dark green top and bottom, and thick.

SIMILAR SPECIES   *E. capitellata* and *E. oblonga,* which have long-fibred stringy bark and globular fruit.

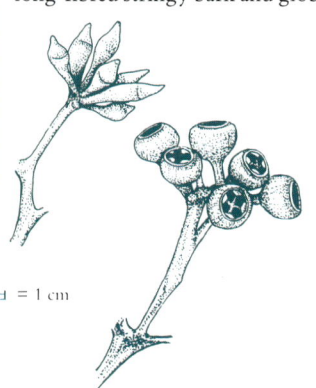

= 1 cm

16

Eucalyptus sparsifolia (new species)

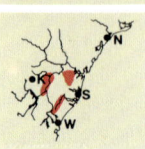

HABITAT   Lower Blue Mountains and in Richmond–Windsor area.

MAIN DISTINGUISHING FEATURES   Stringy bark to smaller branches. Leaves small, narrow and not as thick as other stringybarks.

SIMILAR SPECIES   *E. oblonga* and *E. eugenioides.*

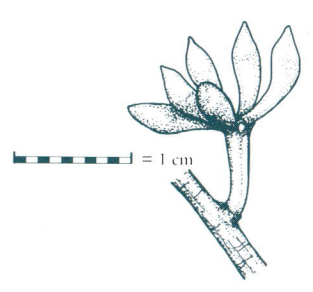
= 1 cm

ASHES

The 'ash' name was originally given to Northern Hemisphere trees of the genus Fraxinus, but many Southern Hemisphere trees are now called ashes. In this book we are of course referring to ashes in the genus Eucalyptus. They are logged extensively for woodchipping. Ash leaves can be distinguished by their venation, which tends to be irregularly spaced and at a shallow angle to the midrib.

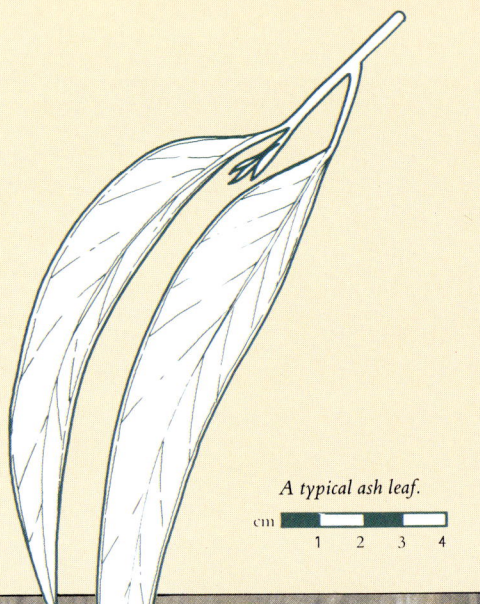

A typical ash leaf.

cm 1 2 3 4

Although called a stringybark, the messmate is an ash. It can be distinguished from true stringybarks by its fruit, which are more egg-shaped.

HABITAT   Cooler mountain areas, Blue Mountains, and to the south.

SOILS   Deep rich loams, especially those derived from Wianamatta shales.

COMMUNITY   Tall open forest.

ASSOCIATED SPECIES   Brown barrel, narrow-leaved peppermint, occasionally with Sydney peppermint on deeper Hawkesbury Sandstone soils.

FORM   Tall, straight trunk; ascending branches.

BARK   Rough, stringy and furrowed but without the long strips and interlacing strands of stringybarks; grey; to smaller branches.

LEAVES   Distinctly oblique base; large; broad, slight curve; glossy green top and bottom.

SIMILAR SPECIES   Can be distinguished from brown barrel, which has smooth upper branches.

INFLORESCENCE   7 to 15; flattened peduncles; late summer.

FRUIT   Egg-shaped.

17

Messmate stringybark
Eucalyptus obliqua

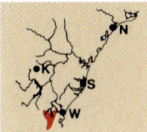

= 1 cm

18

Brown barrel
Eucalyptus fastigata

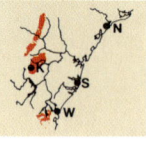

HABITAT   Higher altitudes of the escarpment; high-rainfall areas, such as sheltered parts of upper Blue Mountains, and south Woronora Plateau. Spectacular examples of this tree may be seen on Macquarie Pass.

SOILS   Volcanic; deep loams.

COMMUNITY   Tall open forest.

ASSOCIATED SPECIES   Messmate stringybark, narrow-leaved peppermint.

FORM   Tall, straight trunk; very broad crown. Often seen with long strips of bark hanging from branches.

BARK   Rough, brown; stringy on trunk and larger branches. Upper branches and trunk smooth, white.

LEAVES   Long, curved; oblique; usually same colour top and bottom.

INFLORESCENCE   7 to 15; summer.

FRUIT   Pear-shaped.

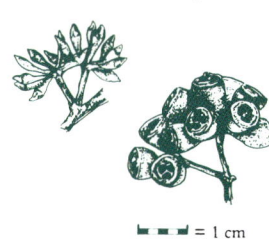

▬ ▬ ▬ = 1 cm

19

Silvertop ash
Eucalyptus sieberi

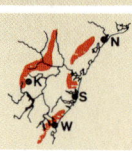

HABITAT   Common on Hawkesbury Sandstones, plateaus of Blue Mountains and Illawarra escarpment; Manly to the Gosford area, which is its northern limit. Does not occur around Sydney.

SOILS   Well-drained, especially Hawkesbury Sandstone.

COMMUNITY   Open forest; woodland.

ASSOCIATED SPECIES   Sydney peppermint, red bloodwood, scribbly gum.

Form   Often a small, bent tree in poor conditions, but can be a medium tree with open canopy on better sites. Low-branching; usually in an erect V-shape.

BARK   Rough; hard; dark; furrowed on trunk and larger branches; grey. Upper branches smooth, white to orange; new stems red.

LEAVES   Sickle-shaped; glossy green; same colour top and bottom; juvenile leaves are blue-grey.

INFLORESCENCE   5 to 15; very white; spring to early summer.

FRUIT   Pear-shaped; wide disc, valves 3; level with rim. Silvertop ash is one of the main species being logged in the Eden area. The young growth is popular with florists.

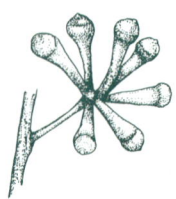

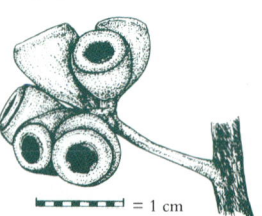

▬ ▬ ▬ ▬ = 1 cm

20

Yertchuck
Eucalyptus
consideniana

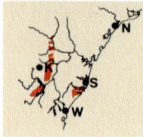

This tree is not common in the Sydney area but may be occasionally found on well-drained soils with silvertop ash, especially on the Woronora Plateau. More common in the Blue Mountains. The bark is more like that of Sydney peppermint; the form is usually stunted, open and straggly.

The juvenile leaves are not bluish, like silvertop ash; and the fruit has 4 valves. Flowering time is late spring.

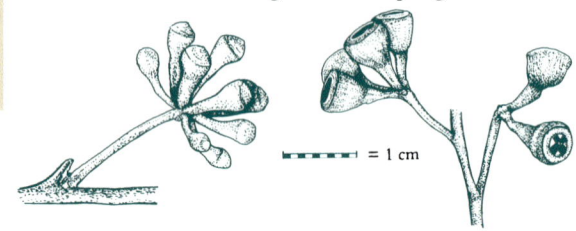

= 1 cm

21

Whipstick
ash
Eucalyptus
multicaulis

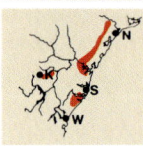

This tree invariably occurs as a mallee and usually grows in association with silvertop ash and red bloodwood. It rarely exceeds 4 m in height, and although having smooth bark it can be easily mistaken for silvertop ash or yertchuck. The flowering time is winter.

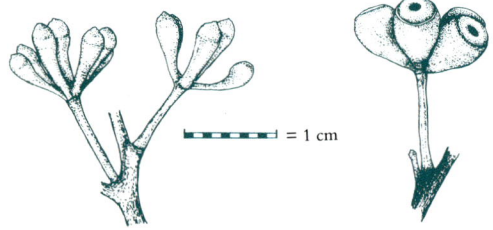

= 1 cm

OTHER ASHES

Species from this group are often very similar in appearance and may tend to hybridize, making positive identification difficult.

22

Faulcon-
bridge
mallee ash
Eucalyptus
burgessiana

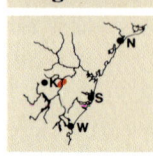

This mallee is similar in appearance and closely related to Port Jackson mallee but occurs only in the Faulconbridge area. Flowering time is summer and generally it has larger leaves and fruit. It is an endangered species.

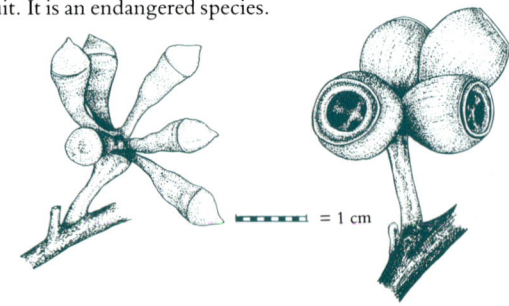

= 1 cm

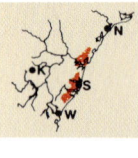

A mallee usually occurring in scrubland or heathland on damp Hawkesbury Sandstones, especially in Ku-ring-gai Chase National Park and Royal National Park. It does not occur more than 25 km from the sea or more than 80 km from Sydney.

The bark is smooth and white to grey.

It can be distinguished by its low growth, many slender stems, and large glossy green sickle-shaped leaves. Young stems are yellow. The flowers are yellow to cream, appearing in spring. It is an endangered species.

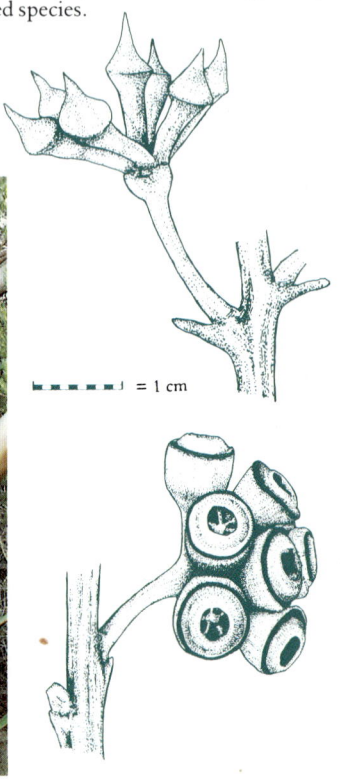

= 1 cm

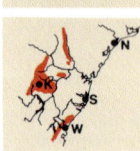

This mallee rarely exceeds 3 m in height and favours shallow Hawkesbury Sandstone soils, often growing in association with red bloodwood, silvertop ash and scribbly gum. It closely resembles Port Jackson mallee but has narrow erect leaves. Flowering extends from summer through autumn.

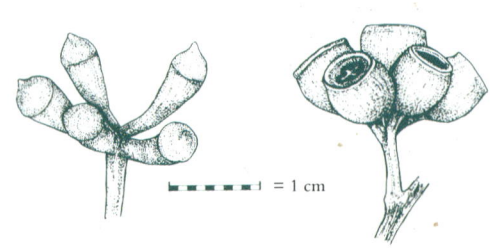

= 1 cm

25

Narrow-leaved mallee
Eucalyptus apiculata

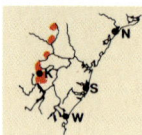

Locally common in the upper Blue Mountains, having a close resemblance to Blue Mountains mallee and occurring in the same habitats. The most consistent distinguishing feature is the narrower leaf (<5 mm). Flowering time summer to autumn.

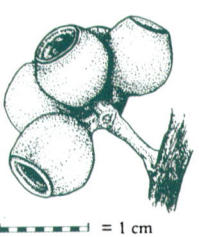

= 1 cm

26

Cliff mallee ash
Eucalyptus cunninghamii
(syn. E. rupicola)

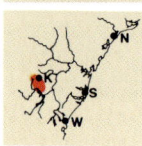

Similar in appearance to narrow-leaved mallee, but smaller (usually no more than 1 m tall) and restricted to cliff edges in the Blue Mountains. The fruits are smaller, and the new leaves are pinkish-grey. Mature leaves are rather densely arranged, with a strong blue tint. Flowering time summer.

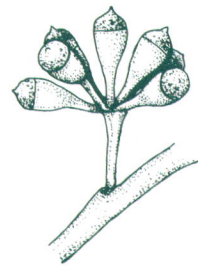

= 1 cm

27

Blue Mountains ash
Eucalyptus oreades

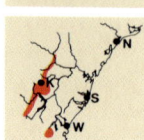

This tree is common in the upper Blue Mountains and is related to yellowtop ash and Budawang ash but differs in that it is generally a well-formed tree. It is found in the Blue Mountains on Hawkesbury Sandstones, in association with silvertop ash, Sydney peppermint and Blaxland's stringybark. The bark is smooth, white to yellow, shedding in long strips. There is a short stocking at the base. The leaves are shiny green and relatively small. Flowering time spring to summer.

= 1 cm

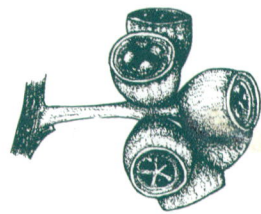

41

28

Budawang ash
Eucalyptus dendromorpha

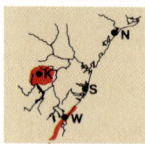

Related to Blue Mountains mallee, but more variable in size, growing consistently as a mallee on the escarpment edge above Wollongong, but occurring as a tree in the Budawang National Park. In the Blue Mountains, both tree and mallee forms may be found. The bark is smooth except for a short brown stocking at the base. The leaves are not as erect as on the Blue Mountains mallee and are dotted with oil glands. Leaves are broad (>15 mm) and sickle-shaped. Flowering time summer. It is an endangered species.

E. obtusiflora has recently been considered to be a variation of *E. dendromorpha* but will appear in previous publications as a separate species.

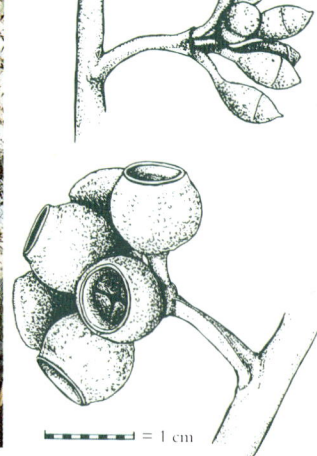

= 1 cm

42

SCRIBBLY GUMS

The bark of these trees is often marked with 'scribbles' created by moth larvae as they burrow beneath the bark.

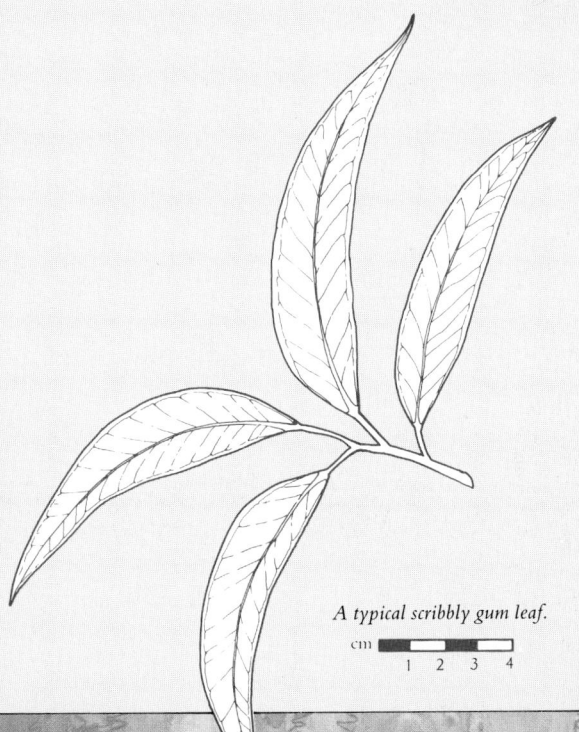

A typical scribbly gum leaf.

cm ▮▮ 1 2 3 4

29

Scribbly gum
Eucalyptus haemastoma

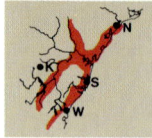

HABITAT Mainly occurring on Hawkesbury Sandstone of the Central Coast and the tableland, but overlapping the habitats of other scribbly gums such as *E. rossii* in the Blue Mountains and *E. racemosa* on the Illawarra escarpment.

SOILS Shallow rocky Hawkesbury Sandstone.

ASSOCIATED SPECIES Other scribbly gums, red bloodwood, silvertop ash.

FORM Tall straight trunk on better sites such as Woronora Dam, but generally with sloping trunk and spreading, twisting branches.

BARK Smooth and white to the ground; when bark is shed grey mottling occurs. Usually marked with 'scribbles' from moth larvae.

LEAVES Leathery, sickle-shaped; dull grey-green top and bottom.

SIMILAR SPECIES *E. racemosa* has narrower leaves and smaller fruit. *E. sclerophylla* has broad, thick, shiny leaves. *E. rossii* has small fruit on a delicate stem.

INFLORESCENCE 7 to 20; white; spring.

FRUIT Pear-shaped; disc broad and red on young fruit.

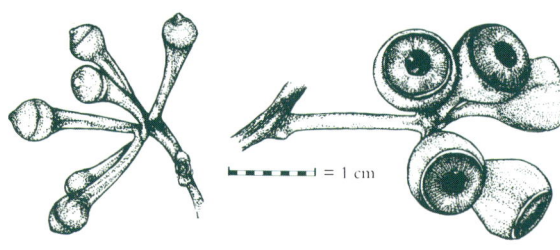

= 1 cm

OTHER SCRIBBLY GUMS

30

Eucalyptus sclerophylla

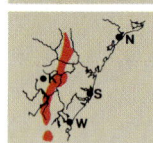

Occurs on the eastern slopes of the Blue Mountains, the Castlereagh area north of Penrith, and the Southern Tableland to Mittagong. Good stands may be seen at Wirrimbirra and Jervis Bay. The fruit are small and often more hemispherical than the other scribbly gums. Flowers in summer.

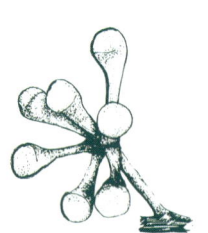
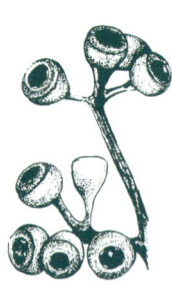

= 1 cm

45

31

Scribbly gum, or snappy gum
Eucalyptus racemosa

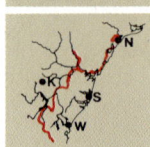

Can be distinguished from *E. haemastoma* by its narrower le (<25 mm) and smaller fruit. It is often a taller, more shapel tree. Flowers in early spring. The species also tolerates a wi variety of soil types and can be found growing on Narrabee series soils at Newport.

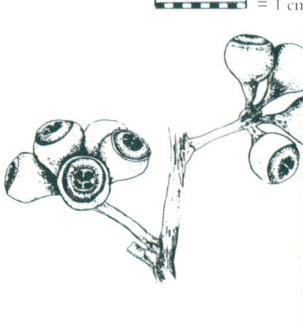

= 1 cm

PEPPERMINTS

Peppermints often grow in association with ashes. Most species have thin, finely fibrous bark. The adult leaves smell strongly of peppermint when crushed, and this oil is commercially extracted. The inflorescences are many-flowered.

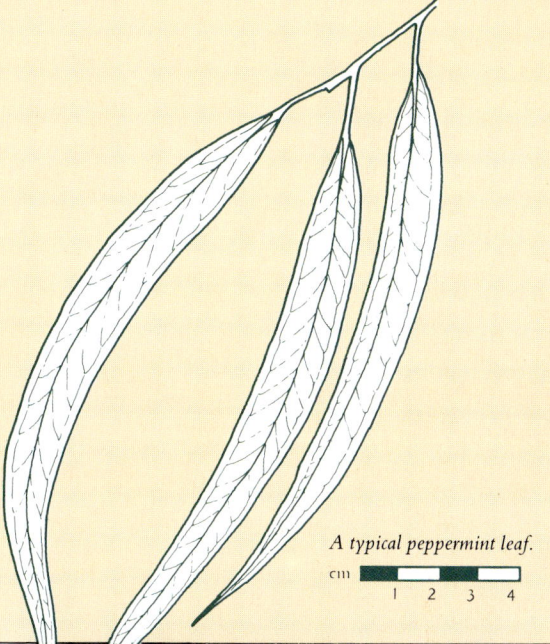

A typical peppermint leaf.

cm

1 2 3 4

32

Narrow-leaved peppermint
Eucalyptus radiata **subspecies** *radiata*

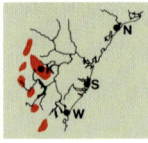

HABITAT   Upper Blue Mountains and Mittagong area, on a wide range of soils.
SOILS   Hawkesbury Sandstone, Wianamatta shales, Narrabeen shales, basalt.
ASSOCIATED SPECIES   Silvertop ash, Sydney peppermint, Budawang ash, messmate stringybark.
FORM   Irregular, from small and bushy to a medium-sized tree, often with low finely textured canopy.
BARK   Grey, rough, fine; persistent to smaller branches.
LEAVES   Thin, narrow, long; green top and bottom; strong peppermint smell; many oil dots. Juvenile foliage is usually plentiful.
INFLORESCENCE   8 to 20; flowering late spring.
FRUIT   Hemispherical.

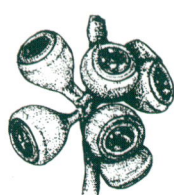

= 1 cm

33

Sydney peppermint
Eucalyptus piperita

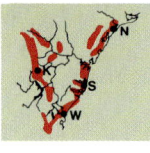

HABITAT   Forests of slopes and ridges with good drainage, common from coast to mountains on sandstone soils.
SOILS   Hawkesbury Sandstone.
ASSOCIATED SPECIES   Silvertop ash, red bloodwood, scribbly gum, grey gum.
FORM   Small to medium tree; trunk usually short, often leaning; canopy low, open and spreading. A more upright tree when growing in forest situations.
BARK   Rough, grey, persistent on trunk and large branches; small branches smooth and off-white to pale grey.
LEAVES   Sickle-shaped, often with an oblique leaf base; strong peppermint smell when crushed; dull bluish-green top and bottom.
INFLORESCENCE   6 to 15; summer.
FRUIT   Egg-shaped; valves enclosed.

Specimens of this species growing at the Helensburgh end of the Royal National Park, and in the Mittagong area produce an urn-shaped capsule. Such specimens were previously described as *E. piperita* subsp *urceolaris,* but are now considered not to have sub-species status.
Originally called E. piperita *subspecies* urceolaris.

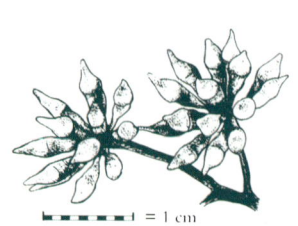
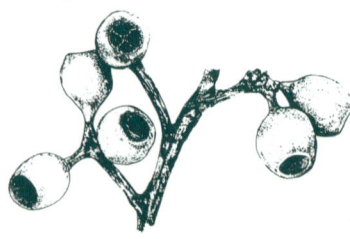

= 1 cm

49

34

River peppermint
Eucalyptus elata

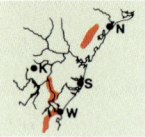

HABITAT   Best specimens follow watercourses and valleys, e.g. upper Kangaroo Valley, Nepean River from Penrith to Menangle, and higher country behind Wollongong. Grows as a smaller specimen on hillsides.

SOILS   Alluvial loams.

ASSOCIATED SPECIES   Woollybutt, narrow-leaved peppermint, gully gum, mountain grey gum.

FORM   Tall well-formed tree, with open, bluish canopy.

BARK   Lower trunk rough, persistent, grey to black, vertically fissured, rather hard; upper trunk and branches shed in long strips to leave smooth white to grey surface.

LEAVES   Bluish-green, narrow, tapering to a long point; strong peppermint smell when crushed.

INFLORESCENCE   7 to 40; spring.

FRUIT   Pear-shaped.

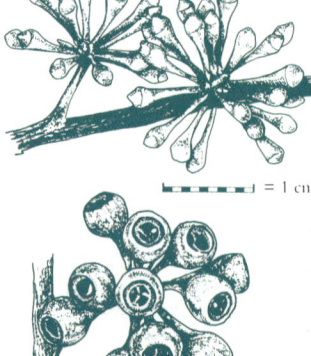

= 1 cm

MAHOGANIES

The rough-barked mahoganies are similar in appearance to some stringybarks, although on close inspection the bark can be seen to lack the long detachable fibres of the stringybarks.

A typical mahogany leaf.

cm

1 2 3 4

35
Bangalay
Eucalyptus botryoides

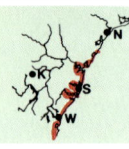

HABITAT   Coastal, especially hind-dunes and creek banks.
SOILS   Sandy.
COMMUNITY   Open forest, woodland.
ASSOCIATED SPECIES   Blackbutt, red bloodwood, Sydney blue gum, white stringybark.
FORM   Small and spreading in hind-dunes, taller, more upright inland.
BARK   Trunk and larger branches covered by thick layer of finely flaky, grey to brown bark; upper branches smooth, cream to grey.
LEAVES   Dark green above, often brown-blotched as a result of damage by brown lace lerp; broad, long, straight, tapering to a narrow point.
SIMILAR SPECIES   Red mahogany has stringier bark, to the smaller branches. Swamp mahogany has flaky, deeply furrowed bark.
INFLORESCENCE   6 to 11 flowers; peduncles long and flattened; summer to autumn.
FRUIT   Usually stalkless, barrel-shaped; valves level with rim.

= 1 cm

36
Eucalyptus saligna– Eucalyptus botryoides **hybrid**

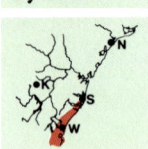

This tree commonly occurs between Port Jackson and Wollongong, especially where disturbance has taken place. Variations that may occur include: length of bark stocking; position of valves; and length of fruit stalk. Hybrid specimens appear to be especially susceptible to brown lace lerp.

Buds and capsules vary in form and size. Depending on habitat, sometimes they may be more like Sydney Blue Gum while at other times more like Bangalay. Collar varies from lower trunk to upper branches.

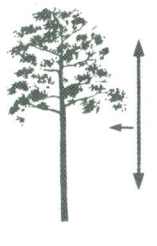

37

Sydney blue gum
Eucalyptus saligna

HABITAT Common on coast and escarpment slopes.
SOILS Moist well-drained soils, especially alluvial sandy loams and podsoils.
COMMUNITY Tall open forest; open forest.
ASSOCIATED SPECIES Blackbutt, grey gum, spotted gum.
FORM Tall and straight, with high narrow branching.
BARK Smooth upper trunk and branches; cream to blue-grey; stocking at base is rough, flaky; brown to grey. Bark shedding in long strips.
LEAVES Green above, may be lerp-damaged; broad, long, tapering to a narrow point.
INFLORESCENCE 7 to 11; flattened peduncles; summer to autumn.
FRUIT Usually with stalks; valves usually above rim, bell- or cylinder-shaped.
SIMILAR SPECIES Flooded gum has an extremely straight, very white trunk. Round-leaved gum has short, broad leaves. Forest red gum has grey-green leaves. Blackbutt has very straight, white to cream main branches in a V-formation.

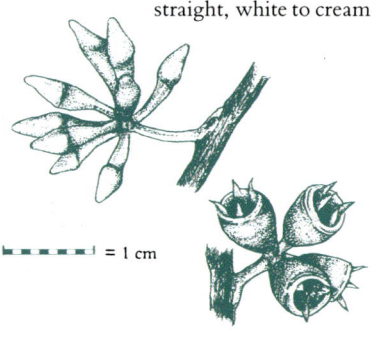

= 1 cm

38

Swamp mahogany
Eucalyptus robusta

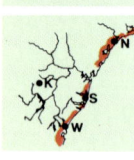

HABITAT Coastal swamps, hind-dunes, lagoons.
SOILS Sand to clay, with high moisture content.
COMMUNITY Open forest.
ASSOCIATED SPECIES Bangalay, forest red gum, red bloodwood, paperbarks, she-oaks.
FORM Straight trunk; spreading branches; crown may be dense.
BARK Rough, persistent to the smaller branches; soft, thick and red-brown; deeply furrowed.
LEAVES Large, thick and shiny; dark green above; tapering to a long point.
INFLORESCENCE 9 to 15; peduncles long, flattened; spring.
FRUIT Cylinder-shaped; valves level with rim.

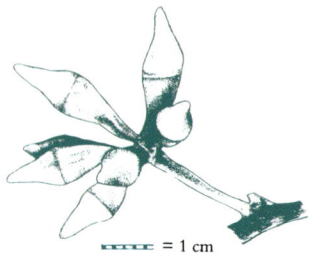

= 1 cm

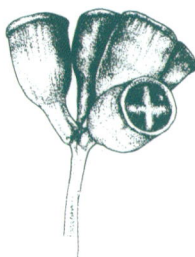

39

Round-leaved gum
Eucalyptus deanei

HABITAT   Sheltered areas, lower slopes and valleys, especial in Blue Mountains.

MAIN FEATURES   Very tall tree; usually only a short brown stocking present. Bark smooth, white with blue-grey patches.

DISTINGUISHING FEATURES   Leaves broader, not as long (noticeable from a distance). Valves don't curve back, as in Sydney blue gum, or curve in, as in flooded gum. Calyptra is not pointed.

ASSOCIATED SPECIES   Sydney blue gum, grey gum.

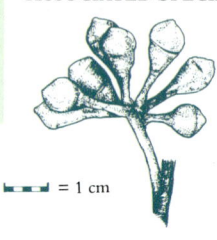

= 1 cm

40

Flooded gum
Eucalyptus grandis

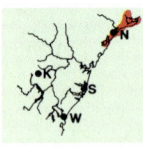

HABITAT   Lower slopes and valleys; flats with deep loams. North of Newcastle.

MAIN FEATURES   Very tall tree; usually there is a short grey stocking; bark smooth, white to blue-grey, powdery.

DISTINGUISHING FEATURES   Valves curve inwards. Fruits may be bluish.

ASSOCIATED SPECIES   Sydney blue gum, blackbutt, red mahogany.

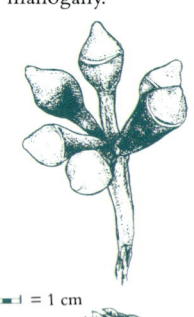

= 1 cm

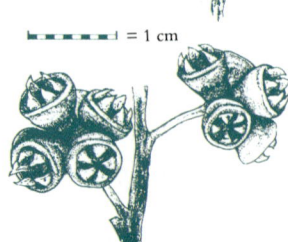

41

Red mahogany
Eucalyptus scias subspecies *scias*
(syn. *E. pellita*)

HABITAT   Coastal plains, especially at moist sites.
MAIN FEATURES   Trunk usually short; crown often broad. Bark rough, thick, furrowed, red–brown, persistent to small branches. Mostly found where soil changes from Hawkesbury Sandstone to shale.
DISTINGUISHING FEATURES   Fruit large, with broad rim; leaves large, dark green, sharply pointed.
ASSOCIATED SPECIES   Forest red gum, red bloodwood, grey ironbark, Sydney blue gum, bangalay.

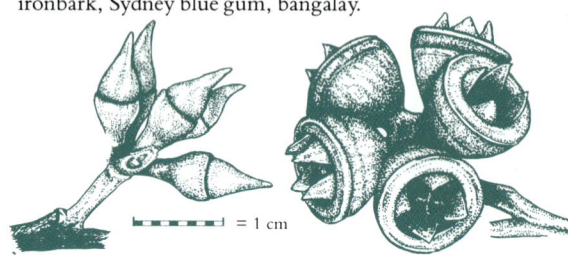

= 1 cm

42

Blue Mountains mahogany
Eucalyptus notabilis

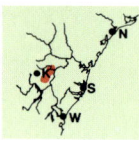

HABITAT   Lower woodland slopes of Blue Mountains.
MAIN FEATURES   Short trunk; spreading crown; bark thick, furrowed, pale grey; only the very smallest branches are smooth; leaves leathery. Very similar in appearance to *E. resinifera*.

= 1 cm

43

Red mahogany
Eucalyptus resinifera subspecies *resinifera*

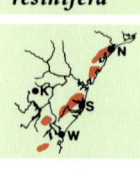

HABITAT   Coastal slopes and flats, especially on clay soils.
MAIN FEATURES   Long straight trunk; bark rough, soft, grey to brown, furrowed; only the smallest branches are smooth.
DISTINGUISHING FEATURES   Operculum very long and horn-shaped.
ASSOCIATED SPECIES   Blackbutt, flooded gum, grey gum.

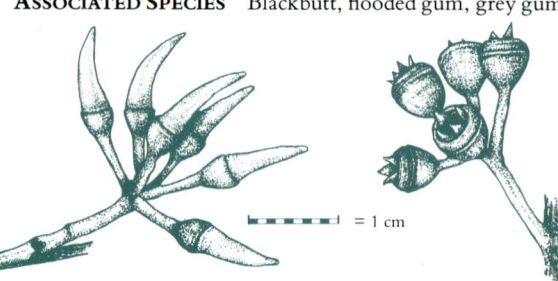

= 1 cm

59

RED GUMS

Named after the colour of their wood, the red gums are an important source of timber for paper, charcoal, and building material. Often characterised by extensive distribution, the distinctive features of the red gums are the smooth, grey and cream mottled bark; and the long calyptra (the cap covering the flower bud) in most species.

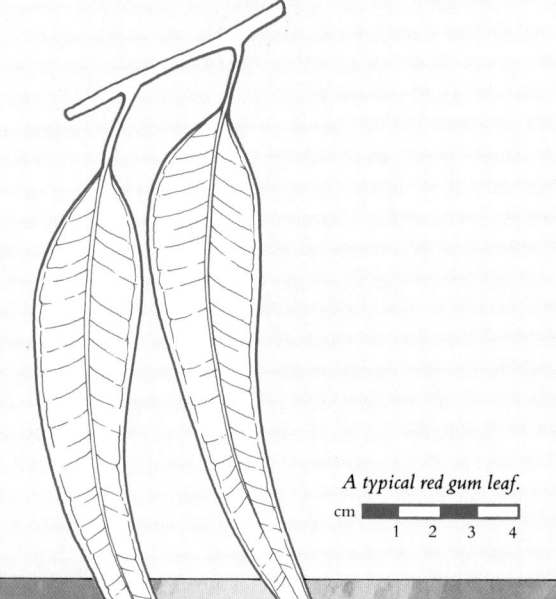

A typical red gum leaf.

cm ▮ ▯ ▮ ▯ ▮
　　1　2　3　4

HABITAT   Very common on coastal plain; escarpment slopes.
SOILS   Clay, occasionally alluvial.
COMMUNITY   Tall open forest; open forest.
ASSOCIATED SPECIES   Grey ironbark, blackbutt,
woollybutt, grey box, red mahogany.
BARK   Usually a dark grey flaky collar at base; rest of trunk
and branches smooth; shedding in patches, leaving an attractive
pattern of white, grey and blue-grey.
FORM   Medium to tall forest tree, with markedly ascending
branches and open canopy.
LEAVES   Long, narrow; slight curve; green to grey-green top
and bottom.
INFLORESCENCE: 7 to 12; spring; the buds are swollen
towards the base.
FRUIT   Globular, with thick, ascending valves.

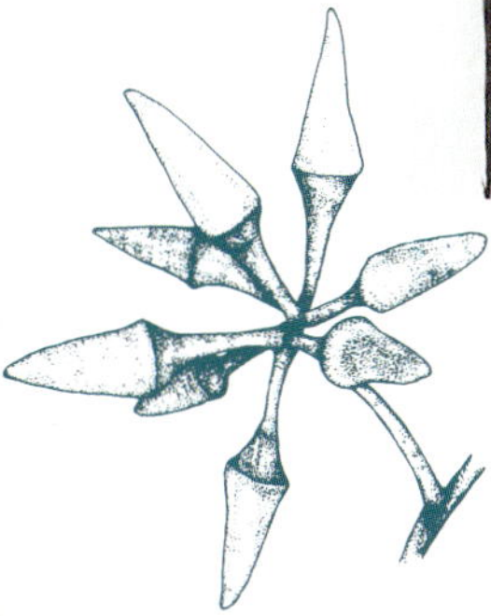

44

**Forest
red gum**
*Eucalyptus
tereticornis*

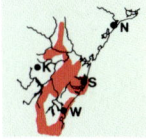

= 1 cm

61

45

Cabbage gum
Eucalyptus amplifolia

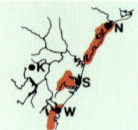

Similar in appearance to forest red gum, but not as commonly occurring, favouring wetter, poorly drained sites, especially along watercourses. The bluish-grey blotching on the bark tends to be darker and contrasts more with the white patches. The buds are more numerous, and are not swollen at the base. Flowers in spring to late summer. The leaves are usually broader and greener.

= 1 cm

46

Parramatta red gum or drooping red gum
Eucalyptus parramat-tensis

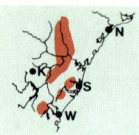

Usually a small woodland tree, with short trunk and open spreading canopy. The buds are smaller and in sevens. Small stands may be found growing in moist clay soils between Picton, Parramatta and Windsor, and along the Putty Road. The bark has a sandpaper texture, and the leaves are small, narrow and sickle-shaped. Summer flowering. *Eucalyptus parramattensis* subspecies *parramattensis* is even less common and is the only red gum in the area with a rounded calyptra.

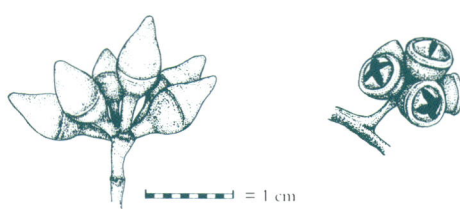

= 1 cm

47

Brittle gum
Eucalyptus michaeliana

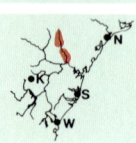

HABITAT Occasionally found along the MacDonald River, north of Wisemans Ferry.
MAIN DISTINGUISHING FEATURES Smooth-barked tree, peeling in plates, giving a mottled whitish-cream-grey appearance.
LEAVES Large, dull green and slightly curved.

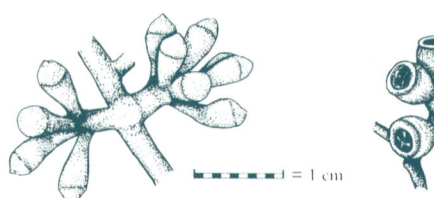

= 1 cm

GREY GUMS

The grey gums were so named by early timber-getters because of the dramatic variation in greys in the bark. The timber is strong and durable, from grey to pale red.

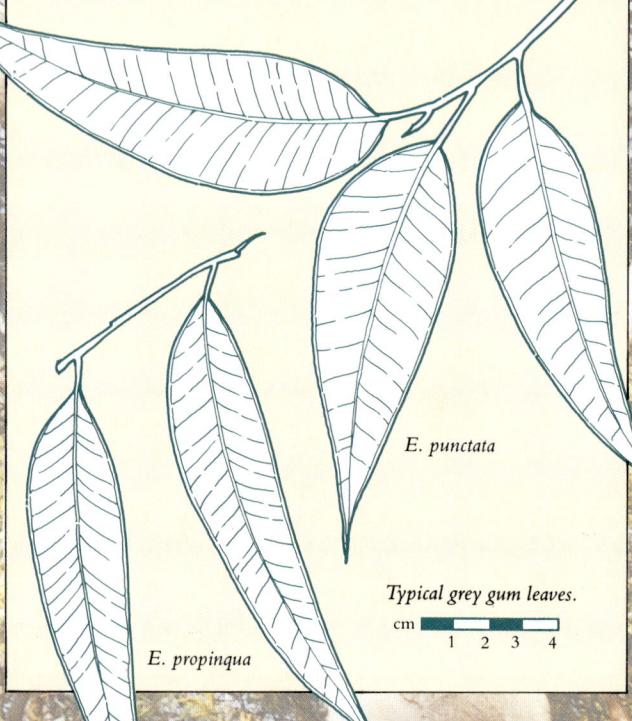

E. punctata

E. propinqua

Typical grey gum leaves.

cm 1 2 3 4

HABITAT   Widespread; coastal hills and ridges; tableland slopes.
SOILS   Often found where shale and sandstone soils merge.
COMMUNITY   Open forest; woodland.
ASSOCIATED SPECIES   Forest red gum, red bloodwood, white stringybark, spotted gum, blackbutt.
FORM   Tall and upright in good conditions; short with spreading crown in poorer soils.
BARK   Smooth, shedding in plates; cream to orange at first, ageing to grey; no collar; often has sandpaper texture.
LEAVES   Long, thin, tapering to a point; dark green above, paler beneath.
INFLORESCENCE   7 to 9 flowers; peduncles angled or flattened; summer to autumn.
FRUIT   Hemispherical; short valves.

Grey gum
Eucalyptus punctata

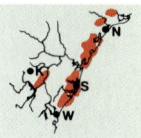

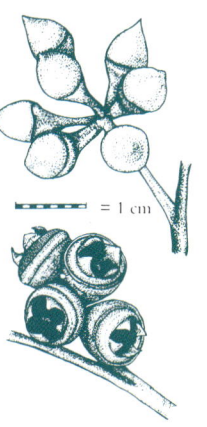

= 1 cm

RELATED SPECIES

Found mainly north of Wyong. Has smaller flowers and fruit than *E. punctata* but produces better timber. Bark is very pale and obscurely patched.

Grey gum
Eucalyptus propinqua

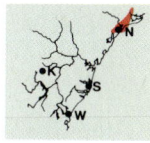

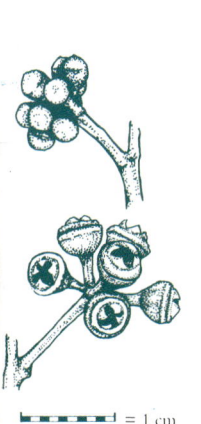

= 1 cm

50

Woollybutt
Eucalyptus longifolia

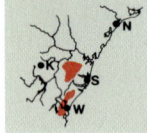

HABITAT   Coastal plains and along watercourses; distributio is patchy.
SOILS   Poorly drained alluvial; shale-derived clay.
COMMUNITY   Open forest.
ASSOCIATED SPECIES   Stringybarks, spotted gum, grey ironbark, coast grey box.
FORM   Medium-sized tree; heavy, irregular branching.
BARK   Rough, grey, persistent to large branches; flaky; close ridged and cracked; smaller branches greenish; shedding in flakes.
LEAVES   Large, pendulous; sickle-shaped; greyish-green bo sides.
INFLORESCENCE   In threes; pendulous; buds large and pinkish-white; spring and autumn.
FRUIT   Bell-shaped.

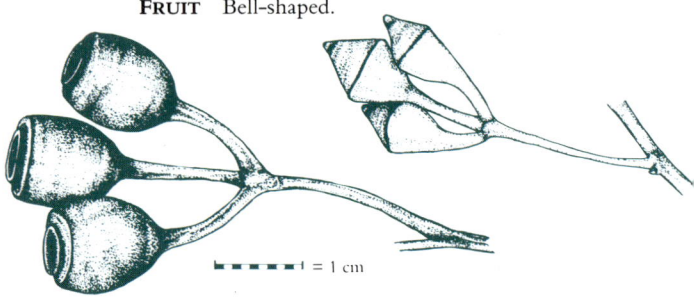

= 1 cm

51

Scaly bark
Eucalyptus squamosa

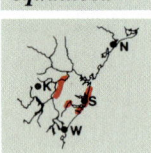

HABITAT   Exposed sandstone ridges in Royal National Park and Broken Bay; foothills of Blue Mountains.
SOILS   Hawkesbury Sandstone, shallow skeletal soils.
COMMUNITY   Woodland; edge of heath.
ASSOCIATED SPECIES   Scribbly gum, silvertop ash, yellowtop ash, narrow-leaved stringy-bark.
FORM   Stunted, straggly small tree; trunk short, leaning and twisted. May appear to be a red bloodwood at first sight, but can be distinguished by the smaller, bluer leaves.
BARK   Rough, grey, persistent to small branches; peeling in thin rectangular flakes.
LEAVES   Small, narrow, sickle-shaped, grey-green top and bottom.
INFLORESCENCE   5 to 12; winter to spring.
FRUIT   Cup-shaped; the valves are exserted, thin and usually recurved.

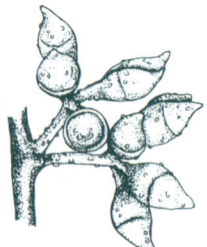

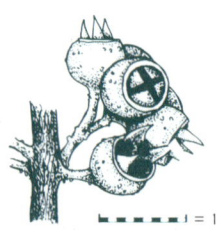

= 1 cm

52

Swamp gum
Eucalyptus ovata

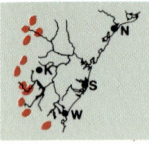

HABITAT   Cooler plateaus and foothills, on poorly drained soils.

SOILS   Wianamatta shale-derived; transitional, from badly drained Hawkesbury Sandstone and shales.

COMMUNITY   Open forest; woodland.

ASSOCIATED SPECIES   Mountain spotted gum, Budawang ash, manna gum.

FORM   From small irregular tree, to medium-sized tree, with sparse crown.

BARK   Persistent rough grey collar; upper trunk and branches smooth; pinkish-cream to white; shedding in long ribbons.

LEAVES   Thick, wide, glossy green top and bottom; edge wavy.

INFLORESCENCE   7; buds appear diamond-shaped; winter.

FRUIT   Cone-shaped.

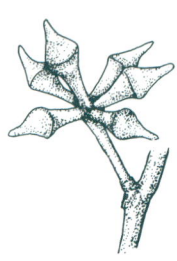
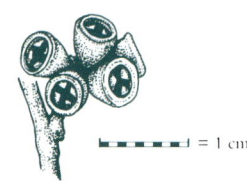

= 1 cm

53

Mountain grey gum
Eucalyptus cypellocarpa

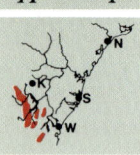

HABITAT   Plateaus and hills of the tablelands; upper Blue Mountains to Mittagong.

SOILS   Narrabeen group on escarpment slopes; deep clay soils.

COMMUNITY   Tall open forest.

ASSOCIATED SPECIES   Brown barrel, coast white box, gully gum, messmate stringybark, yellow stringybark.

FORM   Tall tree with straight trunk and narrow canopy.

BARK   Short flaky grey collar, or smooth to ground; trunk and branches smooth, mottled white and blue-grey shading in strips.

LEAVES   Long, narrow, shiny dark green top and bottom.

SIMILAR SPECIES   Can be distinguished from Blue Mountains ash, which does not have mottled bark, and forest red gum, which has smaller, blue-green leaves.

INFLORESCENCE   7; peduncles flattened; summer.

FRUIT   Cup-shaped, often with ribs.

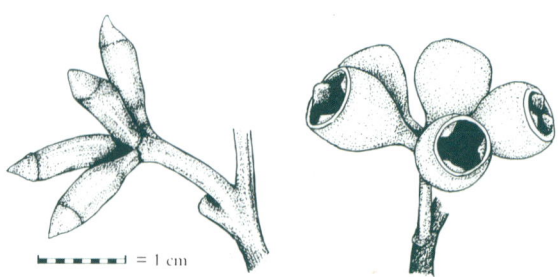

= 1 cm

RIBBON GUMS

Ribbon gums tend to be similar in appearance, having a short bark collar at the base of the trunk and smooth branches with ribbons of peeling bark. In form they tend to be tall and elegant.

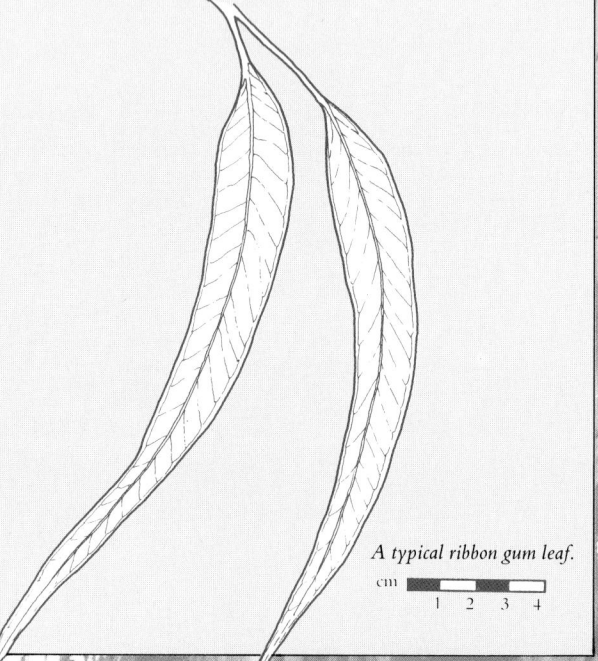

A typical ribbon gum leaf.

cm
1 2 3 4

HABITAT   Escarpment slopes above 300 m altitude; edges of tableland steams.
SOILS   Narrabeen group; clay loams; deep sandy loams.
COMMUNITY   Tall open forest; edges of rainforest.
ASSOCIATED SPECIES   Sydney blue gum, messmate stringybark, yellow stringybark, mountain grey gum.
FORM   Tall forest tree, with long straight trunk and open bluish canopy.
BARK   Hard, rough, dark-brown collar for most of trunk; upper trunk and branches smooth, white, shedding in long strips.
LEAVES   Long, narrow, grey-green top and bottom.
SIMILAR SPECIES   River peppermint often looks very similar but may be distinguished by the peppermint leaf smell. Manna gum has green leaves and softer bark collar.
INFLORESCENCE   5 to 9; late summer.
FRUIT   Ovoid.

54

Gully gum, or blackbutt peppermint
Eucalyptus smithii

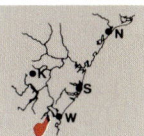

55

Manna gum
Eucalyptus viminalis

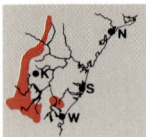

Manna gum may be seen along the Nepean River, especially between Menangle and Camden; also Coxs River and valleys in the Blue Mountains.

The most obvious distinguishing feature is the large quantity of oppositely arranged, stalkless juvenile foliage. The base of the trunk is covered by rough bark to varying extents. The inflorescences are usually 3-flowered, and the fruits are stalkless.

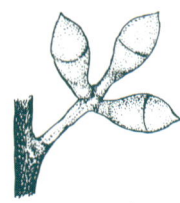

= 1 cm

56

Camden white gum
Eucalyptus benthamii

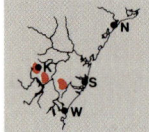

An endangered species, now found occasionally along the banks of the Nepean River between Camden and Wallacia; Lake Burragorang and Kedumba Creek. It is usually a tall, straight-stemmed tree, with smooth white bark and a short flaky collar. The bark sheds in long strips. The buds are small, usually borne in clusters of 4 to 7, and are slightly bluish.

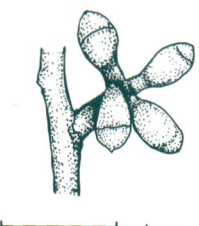

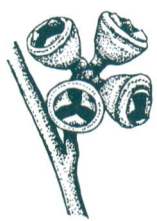

= 1 cm

57

Camden woollybutt, or Paddy's river box
Eucalyptus macarthurii

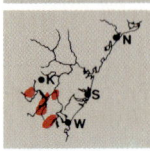

Found in the Blue Mountains, south of Jenolan, and around Moss Vale, Burragorang and Mittagong, usually on poorly drained sites, this tree can be distinguished from manna gum and gully gum by the rough greyish-brown bark, persistent to the larger branches; smaller branches are smooth and grey. The buds are smaller, and the leaves when crushed have a geranium scent.

= 1 cm

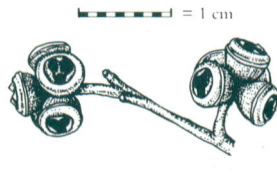

BOXES

Boxes are clearly related to ironbarks but usually have finely textured bark on at least the lower half of the trunk. The branches are generally smooth.

E. quadrangula does not have a typical box leaf. It can be distinguished by its wavy leaf margin.

cm 1 2 3 4

A typical box leaf.

cm 1 2 3 4

BITAT Escarpment slopes where rainfall is high and soil
ep; often close to rainforest. Fine specimens can be seen
.vay down Bulli Pass.

LS Narrabeen group; shale or volcanic soils.

MMUNITY Tall open forest.

OCIATED SPECIES Sydney blue gum, blackbutt, grey
ı, gully gum, yellow stringybark.

RM Tall forest tree with open canopy.

RK Rough, two-toned grey, persistent to small branches;
y fibrous 'box'-like.

VES Long, narrow; green top and bottom; edges are
ularly toothed.

LORESCENCE 4 to 7; late summer.

IT Bell-shaped, stalkless, valves exserted.

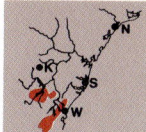

58

Coast white box, or white-topped box
Eucalyptus quadrang-ulata

= 1 cm

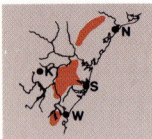

59

Grey box
Eucalyptus moluccana

HABITAT Common on the Cumberland Plain (Parramatta and further west, between Hawkesbury River and Campbelltown) on clay soil and in the Hunter Valley.
SOILS Well-drained alluvial.
COMMUNITY Woodland; open forest.
ASSOCIATED SPECIES Forest red gum, spotted gum, grey gum, grey ironbark, red mahogany.
FORM Straight trunk, to half height of tree; canopy usually V-shaped.
BARK 'Box'-type, thin, grey, persistent collar on half to mo of trunk; branches and upper trunk smooth, light grey, shiny, shedding in long strips.
LEAVES Broad; green top and bottom.
INFLORESCENCE 7; buds diamond-shaped. Late summer to autumn flowering.
FRUIT Barrel-shaped.

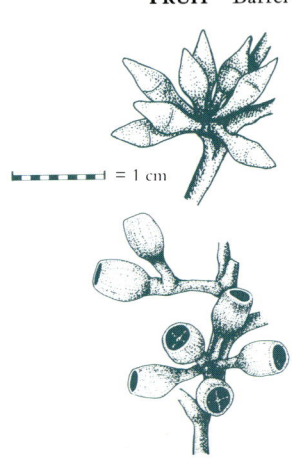

= 1 cm

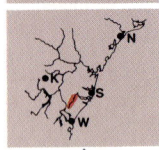

60

Blue box
Eucalyptus baueriana

Blue box may be found on river flats around Richmond, Liverpool and Moss Vale. Especially common along the Georges River in the Moorebank–Milperra area. It is usually seen as a short-trunked woodland tree, with two-toned grey 'box'-type bark persistent to the larger branches. The canopy usually appears bluish and dense because of the large quantitie of juvenile foliage.

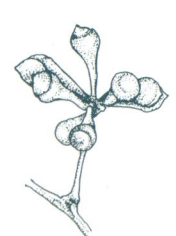

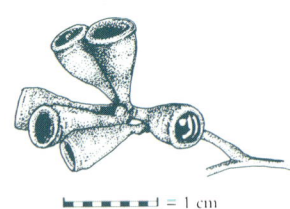

= 1 cm

61

Coast grey box
Eucalyptus bosistoana

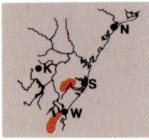

Coast grey box is the largest of the boxes and is found more frequently in open forest than in woodland.

HABITAT   River flats in Liverpool and Illawarra; Grose and Burragorang Valleys; becoming more common south of Wollongong.

SOILS   Alluvial loams.

COMMUNITY   Open forest.

ASSOCIATED SPECIES   River red gum, small-leaved stringybark, woollybutt, grey ironbark.

FORM   Tall forest tree with straight trunk; canopy dense and dark green.

BARK   Grey to fawn 'box'-type collar, persistent on lower trunk; upper trunk and branches smooth, grey and white, shedding in long strips.

LEAVES   Long, narrow, thin, green top and bottom.

INFLORESCENCE   3 to 7; buds egg-shaped; summer flowering.

FRUIT   Cup-shaped.

This species may easily be mistaken for forest red gum with which it is often found. A close examination of the bark, buds and capsules will often be necessary.

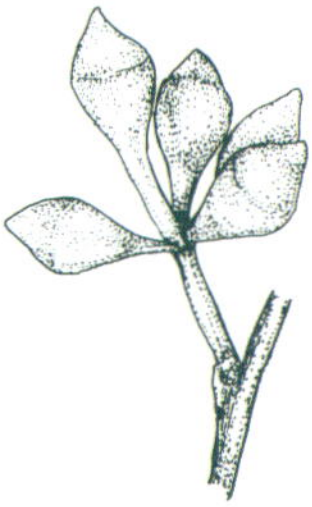

1 = 1 cm

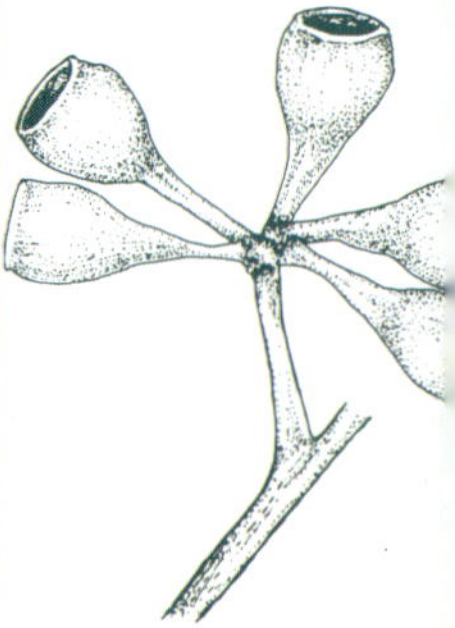

IRONBARKS

Most ironbarks have very hard, thick bark, often deeply fissured longitudinally. In most species the rough bark extends to the larger branches. Their canopies tend to be fairly dense.

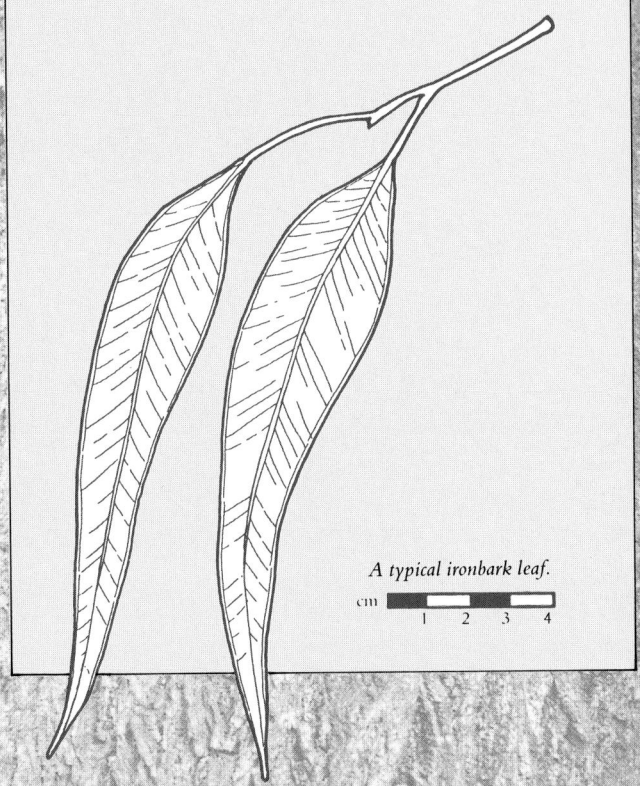

A typical ironbark leaf.

cm 1 2 3 4

62

Grey ironbark
Eucalyptus paniculata subspecies *paniculata*

HABITAT   Illawarra and Blue Mountains escarpment slopes and foothills; coastal plain; valleys.
SOILS   Sandy loams; clays.
COMMUNITY   Tall open forest; open forest.
ASSOCIATED SPECIES   Blackbutt, forest red gum, woollybutt, coast grey box, spotted gum, grey gum.
FORM   Tall forest tree, with long straight trunk and dense canopy.
BARK   Hard, light grey to black ironbark, persistent to smaller branches; deeply furrowed.
LEAVES   Long, narrow; dark green; paler beneath.
INFLORESCENCE   In sevens; winter to spring.
FRUIT   Large, thick, pear–shaped.

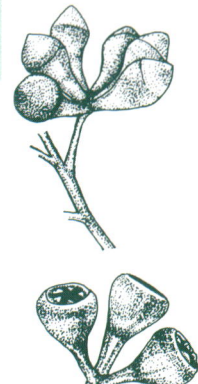

= 1 cm

OTHER IRONBARKS

63

Narrow-leaved ironbark
Eucalyptus crebra

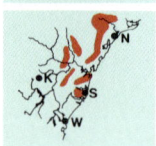

HABITAT   Ridges, plateaus, undulating plains; on sandy and clay loams in woodland; foothills of Blue Mountains; on Wianamatta shale in Picton area.
MAIN IDENTIFYING FEATURES   Dark grey, deeply furrowed ironbark with tall trunk and sparse, straggly canopy.
DISTINGUISHING FEATURES   The fruit are small, and the leaves are variable, but generally narrow, and dull grey-green top and bottom.
ASSOCIATED SPECIES   Other ironbarks, forest red gum, grey box, grey gum.

= 1 cm

64

Eucalyptus beyeriana (syn. E. beyeri)

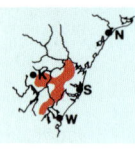

This species may be easily mistaken for *E. crebra*, the only othe fine-leaved ironbark in the Sydney region. The leaves of *E. beyeriana* are darker on the upper surface, whereas those of *E. crebra* are the same colour top and bottom. It may be found the western suburbs, lower Blue Mountains, and along the freeway to Mittagong.

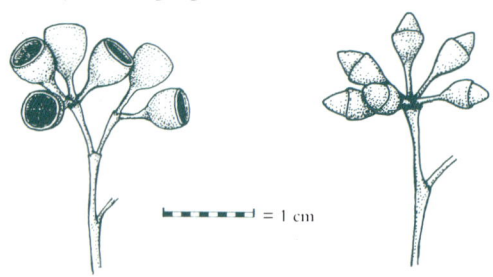

= 1 cm

65

Broad-leaved ironbark Eucalyptus fibrosa

HABITAT   Drier coastal plains and hills; good specimens can be seen in the Windsor–Richmond area.

MAIN IDENTIFYING FEATURES   Grey-black ironbark, somewhat flaky, persistent to smaller branches; straight trunk and spreading canopy.

DISTINGUISHING FEATURES   Bark is deeply fissured, but not as hard as most ironbarks. The leaves are grey-green top and bottom. The calyptra is very long and sharply tapering.

ASSOCIATED SPECIES   Other ironbarks, spotted gum, white stringybark.

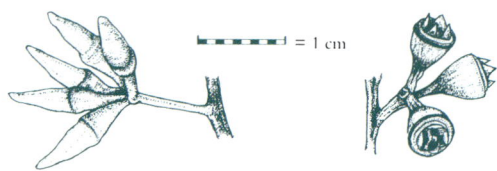

= 1 cm

66

Red ironbark Eucalyptus sideroxylon

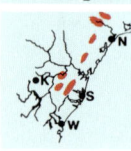

HABITAT   Plains and gradual slopes; on gravels and clays; Picton to Liverpool, and Hunter Valley.

MAIN IDENTIFYING FEATURES   Hard, very dark ironbark, persistent to smaller branches; usually short trunk and dense spreading canopy.

DISTINGUISHING FEATURES   The bark is darker than other ironbarks; the inflorescences are pendulous, and the flowers may be white, pink, red or yellow.

ASSOCIATED SPECIES   Other ironbarks, *E. tricarpa* (formerl *E. sideroxylon* subspecies *tricarpa*) has longer flowers in 3s and, although not native to the area, is often planted for ornamental purposes.

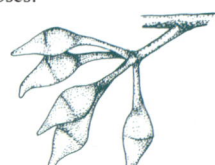
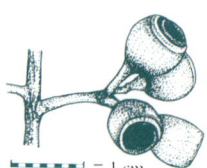

= 1 cm

67

Northern grey ironbark
Eucalyptus siderophloia

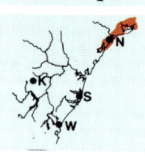

HABITAT Commonly occurring along the MacDonald River, Colo River, and northwards.

MAIN IDENTIFYING FEATURES Grey ironbark, with long grey-green leaves tapering to a fine, recurved point.

SIMILAR SPECIES *E. crebra* and *E. beyeriana* have narrower leaves. *E. sideroxylon* has darker bark.

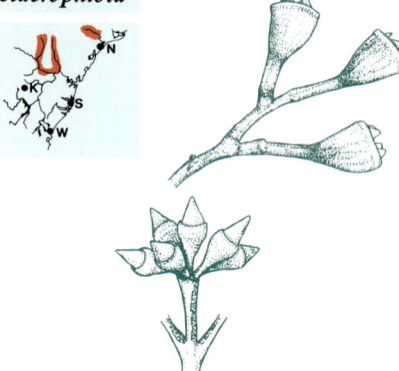

= 1 cm

TALLOWWOOD

68

Tallowwood
Eucalyptus microcorys

The southernmost limit of tallowwood is Dora Creek, but as it is so commonly planted in the Sydney area, one might assume that it is a local species. It is one of the best hardwood timbers of New South Wales.

HABITAT Slopes and ridges, especially close to rainforest.

SOILS Moist clays and sandy loams.

COMMUNITY Tall open forest.

ASSOCIATED SPECIES Sydney blue gum, blackbutt, white-topped box, flooded gum.

FORM Tall forest tree, with tall straight trunk and dense, spreading canopy.

BARK Rough, soft and fibrous; brown; persistent to smaller branches.

LEAVES Thin, long, tapering to sharp point; green, paler beneath; edges are finely scalloped.

SIMILAR SPECIES Often found growing with *E. acmenoides*, which appears very similar in form. A close examination of buds and fruit may be necessary to make the distinction.

INFLORESCENCE 7 to 9; buds cream, with short operculum.

FRUIT Pear-shaped.

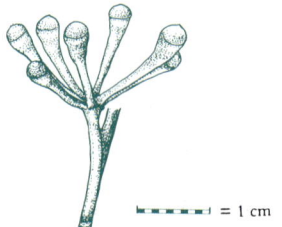

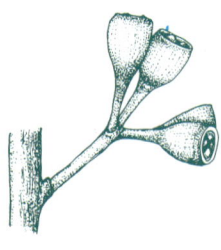

= 1 cm

85

FURTHER READING

As the descriptions in this field guide are brief, further confirmation could be obtained from the following books:

Flora of Australia Vol. 19: Myrtaceae, Eucalyptus, Angophora (Australian Government Publishing Service, Canberra, 1988).

Flora of NSW Vol. 2 edited by G. Harden (NSW University Press, Sydney, 1991).

Flora of the Sydney Region by N.C.W. Beadle et al. (Reed, Sydney, 1982, 3rd edition).

Forest Trees of Australia by D.J. Boland et al. (Thomas Nelson and CSIRO, Melbourne, 1984, 4th edition).

Native Plants of the Sydney District by A. Fairley and P. Moore (Kangaroo Press, Sydney, 1989).

GLOSSARY

ALLUVIAL SOIL:	Moved and deposited by water.
ASSOCIATION:	A community of plants having several dominant members, which have similarities in structure.
BOLE:	The trunk, from the ground to the first branches.
CALYPTRA:	The 'cap' covering a flower bud.
CALYX:	The sepals, which form the outer whorl of the flower.
CALYX TUBE:	Formed when the sepals are joined.
COMMUNITY:	A naturally occurring group of plants.
COROLLA:	The petals of a flower.
DECORTICATE:	The annual shedding of bark.
ENDEMIC:	Growing only in a particular area.
EPICORMIC:	Growth arising from buds beneath the bark, usually after the bark has been wounded.
EXSERTED:	Thrust forth, or thrust out.
GRANITE:	Volcanic rock.
HABITAT:	The place where the plant is growing.
INFLORESCENCE:	A group of flowers, and the structure that supports them.
INTERGRADE:	Two similar species growing in the same habitat and showing some shared features.
INTRAMARGINAL VEIN:	A vein that roughly follows the leaf edge.
LIGNOTUBER:	Swollen tissue at the base of the tree, which is able to produce new shoots.
MALLEE:	Multi-stemmed tree; the stems arise from lignotubers.
OPERCULUM:	The calyptra of a eucalypt flower.
PEDICEL:	The stalk of a single flower.
PEDUNCLE:	The stem that supports an inflorescence.
PERSISTENT BARK:	Bark that does not decorticate.
PLAINS:	Flat terrain.
RIDGE:	Top of an escarpment or range.
TALUS SLOPE:	Escarpment slope, usually covered with eroded material.
TESSELLATED BARK:	Bark that is divided into square or rectangular chunks.
UNDERSTOREY:	Plants growing beneath the trees.
UNDULATING:	Having low hills and shallow valleys.
VALVES:	The sections that divide to release seed from the capsule.

INDEX
TO TREE REFERENCE NUMBERS